T0275801

Most natural populations intermittently experience extremely stressful conditions. This book outlines how such conditions can influence evolutionary change. Extreme conditions cause periods of intense selection. They may enhance fitness differences among genotypes and lead to associations between heterozygosity and fitness. Both phenotypic variation and genetic variation can increase under stressful conditions, particularly for traits that normally show a low level of variation. As a consequence, rapid evolution is possible when conditions are extreme. The fossil record indicates periods of rapid evolutionary diversification associated with drastic environmental changes, especially where productivity is high. Uncommon conditions that are stressful can have a major impact on the evolution of life history traits and lead to tradeoffs between environments. The evolutionary effects of extreme changes are important in planning conservation strategies for the long-term persistence of species.

Extreme environmental change and evolution

Extreme environmental change and evolution

ARY A. HOFFMANN

School of Genetics and Human Variation
La Trobe University, Australia

PETER A. PARSONS

School of Genetics and Human Variation
La Trobe University, Australia

CAMBRIDGE UNIVERSITY PRESS
Cambridge, New York, Melbourne, Madrid, Cape Town, Singapore,
São Paulo, Delhi, Dubai, Tokyo

Cambridge University Press
The Edinburgh Building, Cambridge CB2 8RU, UK

Published in the United States of America by Cambridge University Press, New York

www.cambridge.org
Information on this title: www.cambridge.org/9780521446594

First published 1997

A catalogue record for this publication is available from the British Library

Library of Congress Cataloguing in Publication data

Hoffmann, Ary A.
    Extreme environmental change and evolution / Ary A. Hoffmann,
Peter A. Parsons
        p.   cm.
    Includes bibliographical references and index.
    ISBN 0 521 44107 2. ISBN 0 521 44659 7 (pbk.)
    1. Adaptation (Biology)   2. Extreme environments.   3. Evolutionary
genetics.   4. Evolution (Biology)   I. Parsons, P. A. (Peter Angas),
1933–   .   II. Title.
QH546.H63   1997
576.8′4–dc21   96–46901   CIP

ISBN 978-0-521-44107-0 Hardback
ISBN 978-0-521-44659-4 Paperback

Transferred to digital printing 2010

Contents

Preface

This book arose out of our perception that extreme environments have an important impact on evolutionary change. This perception stems from a wide ranging literature review, and from our own field and laboratory work. We present an approach to this area based on diverse procedures used in evolutionary studies. These range from detailed studies of evolutionary processes within populations, to optimality approaches that consider the outcome of evolution, and patterns of evolutionary change in the fossil record. We do not provide a detailed literature survey of these diverse areas. Instead, we focus on those studies and concepts that link evolutionary changes to extreme conditions. Our intent is to make the material accessible to a wide audience rather than to be comprehensive. Some areas of ongoing research have not been included in this book. In particular, we have focussed on environmental extremes associated with abiotic factors, rather than the effects of biotic factors such as pests and diseases. However, interactions among abiotic and biotic factors are noted in several places.

We feel that a book in this area is warranted because extreme environments are being largely ignored in the current evolutionary literature. We perceive three main reasons for this neglect. Firstly, the major advances in molecular techniques in the last decade have led to a concerted focus on the outcome of evolutionary changes, rather than on the process itself. It is now possible to trace the DNA changes that have occurred in an increasingly larger number of genes. These studies provide a detailed description of how evolution has occurred and the relative importance of genetic drift and selection in shaping these changes, but they do not provide an understanding of how the environment and genome interact in driving adaptive evolution. Secondly, there has recently been a renewed interest in the quantitative analysis of selection in natural populations. These analyses often assume that the amount of genetic variation in traits and the ways that traits interact genetically is relatively constant over time, irrespective of the environment. Consequently, the emphasis in these studies is on a statistical description of variation in traits, rather than the interaction between traits and the environment. Thirdly, Charles Darwin argued that competition is an important force in shaping evolutionary change, and this factor has been emphasized in many evolutionary studies. However, we suggest that environmental effects and perturbations are common and often override such biotic effects.

Our aim is to provide a book that is both useful in teaching and of general interest. Much of the material has previously been presented to a third year undergraduate class in evolutionary biology. Several chapters have also been used at the Honours and postgraduate levels. We also hope that this book will generate further interest in the role of extreme environments in evolution. Most of the answers to the important questions in this area remain unclear. This is because experiments on the environmental edge close to extinctions are more difficult to undertake than those using benign environments. Obtaining these answers is particularly pertinent in light of the increasingly extreme environments that activities of our own species are imposing on other organisms.

A number of people have commented on earlier drafts of the chapters. In particular, we are grateful to Kathryn Bays, David Berrigan, Mark Blows, Anthony Hallam, Miriam Hercus, Nicole Jenkins, Louise Parsons, Carla Sgrò, and Richard Woods. We also thank our editor at Cambridge University Press, Tim Benton, for numerous comments that markedly improved the contents and focus of the book.

CHAPTER 1

Introduction

It should be obvious that the world can be a hostile place for species of plants and animals including our own, because environments are constantly changing. Floods, fires, storms, volcanic eruptions, earthquakes and sudden releases of chemicals highlight the extent to which environments can be made rapidly inhospitable for organisms by catastrophic events. Many less dramatic environmental changes can also be devastating. Periods of low rainfall can result in severe droughts that decimate populations of animals and plants over wide areas. Changes in the temperature of oceans can alter the abundance of plankton so that starvation occurs in marine animals higher up the food chain relying on plankton for food. Encroaching salinity from rising groundwater levels can lead to the extinction of entire plant communities as well as the animals that depend on these communities for food and shelter.

Human activities are implicated in many of the recent catastrophic environmental changes. At a localized level, there are numerous examples of oil and chemical spills killing large numbers of birds and aquatic organisms. On a larger scale, forest fragmentation can cause many extinctions. Surface and soil temperatures become elevated and humidity levels fall in fragmented forests. This can lead to the extinction of many animal and plant populations sensitive to low humidity and warm temperatures, as has been demonstrated in the deforested regions of Amazonian rainforests. The effects of air pollution are becoming increasingly evident in Europe and North America, where interactions among several pollutants including ozone, sulphur dioxide and nitrogen oxides are severely damaging forests. As industrial activities and land clearing continue to expand, we can expect an acceleration of this decline and fragmentation of forests and other ecosystems.

It is less widely appreciated that natural environments have always been hostile for organisms, at least sporadically, even in the absence of human interference. White (1993) has amassed much evidence, especially in herbivorous animals, that the abundance of organisms is often determined by a relative shortage of resources for the young, especially sources of nitrogen that animals need for the production of proteins. Most organisms have the capacity to produce many offspring, but only rarely are sufficient resources available for large numbers of these offspring to survive and reproduce. Because of this, White (1993) has argued that most animals are likely to encounter environments lacking in nitrogen and other resources.

The instability of natural populations and environments has been emphasized by many other ecologists. As natural populations of animals continue to be monitored for longer periods of time, it is becoming apparent that many populations change drastically in size over long intervals even though they may appear to be relatively stable over shorter periods spanning a few years (Pimm, 1991). These changes are likely to be triggered by environmental conditions. In their classic book, *The Distribution and Abundance of Animals*, Andrewartha & Birch (1954) pointed out that environmental factors, particularly those related to climate, are changing continuously. This often leads to marked fluctuations in the abundance of species, especially when the supply of food is altered.

Environmental changes are not a recent phenomenon, but have occurred throughout geological time. Organisms have always been exposed to fluctuations in factors such as temperature, humidity, the composition of the atmosphere and sea level. When drastic changes in environmental conditions occur, large numbers of species can become extinct. The fossil record shows five periods of mass extinction when a substantial fraction of species disappeared, interspersed with other short periods when extinction levels were lower but still substantially above levels that are normally observed. During periods of mass extinctions, unrelated organisms have become extinct in many parts of the world. This implies that major global environmental changes are responsible for such events (Raup, 1991). In addition, less drastic changes have occurred throughout geological time. For instance in southwestern USA, there have been repeated pulses of increased precipitation in the period between 20 000 and 15 000 years ago (Allen & Anderson, 1993). Large pulses occurred every 2000 years, interspersed with smaller pulses separated by around 200 years.

Because drastic environmental changes are common and have been occurring for a long time, we need to consider their impact on the evolution of life. How do such changes influence evolutionary processes? Do they enhance or retard the rate of evolution? What features of organisms affect their ability to evolve in response to the appearance of extreme conditions?

This book represents an attempt to consider and provide answers to these questions. It is our contention that much of evolutionary change is triggered by extreme conditions. As a byproduct of this task, we aim to provide insights into the potential of organisms to evolve in response to the escalating environmental changes being generated by human activities.

Before considering the impact of drastic changes on evolution, we briefly examine the direct and indirect effects such changes have on organisms. In the next section, we define the types of environmental changes and their effects. We then discuss ways of measuring these effects, particularly by focussing on the way organisms perform under a range of conditions. This is followed by a final section looking at four important evolutionary consequences of extreme conditions that form the basis of the ensuing chapters.

Direct and indirect effects of environmental change

In this section, we examine ways environmental changes influence organisms. As we will see, effects of these changes can be direct when organisms are unable to cope with new conditions. In addition, there are numerous indirect effects because of interactions among organisms.

Direct effects occur when organisms are influenced immediately by changes in physical or 'abiotic' features of the environment such as temperature, humidity, rainfall or toxins. These can impose unfavourable conditions on organisms by increasing mortality rates, or by reducing the availability of food and other resources.

Changes in abiotic features that have drastic effects on the survival and/or reproduction of organisms are commonly referred to as 'stresses', and organisms that experience these effects are often said to be 'stressed'. The effects of a stress can be measured by its influence on the 'fitness' of an organism, which is a measure of the success an organism has in leaving descendants. In fact, stress has been defined as 'any environmental change that reduces the fitness of an organism' (Koehn & Bayne, 1989). In the context of this book, we should probably amend this definition to 'any environmental change that drastically reduces the fitness of an organism' because, as implied above, most individuals do not attain the maximum reproductive and survival rates that they are capable of under optimal conditions. We therefore use the term 'stress' only to refer to extreme situations infrequently encountered by organisms.

Extremes of abiotic factors can cause death when the physiological tolerance limits of organisms are exceeded. These limits vary enormously for different organisms. A short spell at 5°C might result in the death of a tropical species that normally experiences relatively warm temperatures throughout the year, whereas several days well below 0°C might be required before death occurs in temperate species adapted to seasonal fluctuations. Physiological limits also depend on the developmental stage an organism has reached. Juveniles tend to be more susceptible than mature adults, so that mortality is often much higher in young age classes when environmental extremes are encountered. Stresses may impose less severe direct effects on organisms, such as decreasing an organism's ability to produce progeny, increasing the time an organism takes to develop and become sexually mature, or decreasing its growth rate. These effects are all manifestations of a fall in fitness.

There are many recorded cases where environmental changes have directly influenced organisms. For instance, during the 'El Niño event' of 1982–3, there was movement of water from the tropics into more southern areas of the Pacific Ocean, mainly along the coast of Ecuador and Peru. As a consequence, water temperatures increased and the availability of nutrients decreased. This led to high death rates in some groups of organisms as a direct consequence of the altered environmental conditions (Glynn, 1988); many corals died from high temperature stress, and in some cases entire coral reefs became extinct. In addition, there was a decline in populations of benthic algae which live at the bottom of the sea, because these algae could not withstand the high water temperatures and low nutrient levels.

In many cases, climatic changes cause resources essential to organisms to become limiting in an environment. The availability of soil nutrients or soil water can become restrictive for plants, reducing their growth rate and seed production. As a consequence, food may be abnormally scarce for herbivorous animals, preventing them from acquiring sufficient energy and nutrients. Direct environmental effects on one organism therefore can lead to effects on other organisms. A particularly well-documented case is the effect of drought on populations of the checker spot butterfly, *Euphydryas editha*, in California, in 1975–7 (Ehrlich *et al.*, 1980). Several populations became extinct during the drought period, while others were drastically reduced in size. The host plants of *Euphydryas* butterflies died off rapidly at times of drought stress. This prevented the successful development of butterfly larvae, resulting in their death (Weiss *et al.*, 1987). The El Niño event of 1982–3 provides another example. A decline in plankton during this event led to a food shortage for fish (Glynn, 1988). As fish numbers declined, a food shortage developed for sea birds, fur seals, and sea lions that were dependant on fish. As a consequence, many of these groups could not breed successfully, and populations declined. Finally, there are several documented cases of severe drought conditions causing a decline in mammal populations because of food shortages. In Australia, populations of kangaroos and other marsupials often undergo drastic reductions in size following persistent hot and dry conditions that kill food plants (e.g., Caughley, Grigg & Smith, 1985).

We have so far focussed on climatic changes, and it should be emphasized that other stresses can also have a range of effects. For instance, chemicals can influence both the survival and reproduction of organisms. Heavy metals that accumulate in soils as a consequence of mining are often toxic to plants and prevent many species growing. Water pollutants can kill populations of fish and invertebrates, as highlighted by the effects of disasters such as the breaking up of oil tankers and the release of toxic waste into river systems. Apart from their immediate effects, chemicals have many delayed effects on organisms. For instance, oil spills can result in surviving fish and invertebrates producing offspring with a high incidence of abnormalities. In birds of prey, pesticides can cause reproductive failure when they eventually accumulate to levels that cause weakening of the egg shells.

Apart from acting as stresses, environmental changes also have indirect effects on organisms by influencing 'biotic' or biological features of the environment, namely competition, predation and disease. Taking competition as an example, consider two species that are in direct competition with each other for the same food source along an environmental gradient. A gradient may consist of a change in any abiotic variable such as mean or maximum temperature over a geographical area. Assume that species A has a greater competitive ability than species B at one end of the gradient but the reverse is true at the opposite end. In the absence of a competitor, both species may be capable of surviving along the entire gradient, but competition determines their relative distributions. This may shift as the environmental gradient changes. While the distribution shift would be induced by an abiotic factor such as temperature, it is mediated through competition.

There are several laboratory studies documenting switches in the competitive superiority of two species between environments, as envisaged in this simple case. Arthur (1987) described competition experiments between two well-studied *Drosophila* species, *D. melanogaster* and *D. simulans*, where the former species is at an advantage when flies are raised at 25°C but *D. simulans* is at an advantage when flies are raised at 15°C. Although the nature of the competitive advantage is not understood, 25°C is considered close to optimal for *D. melanogaster* cultured in the absence of *D. simulans*, whereas optimal performance in *D. simulans* occurs at a somewhat lower temperature. Competitive ability in this case can be predicted by the performance of the species in non-competitive situations. Switches in competitive ability with environmental conditions have also been found in laboratory competition experiments between other *Drosophila* species and between different species of flour beetles (see Arthur, 1987).

In the field, there are cases where the relative competitive ability of related species switches with environmental conditions. Ford (1982) reviewed the interactions between two shore barnacles, *Balanus balanoides* and *Chthamalus stellatus*. The former species is adapted to colder conditions, and becomes rare when temperatures in the warmest month exceed 18°C and temperatures in the coldest month exceed 8.9°C. On the other hand, *C. stellatus* favours warmer waters. The two species only occur together in Britain and northern France. Between the 1930s and 1950s, *B. balanoides* was replaced by *C. stellatus* in many areas in Britain, corresponding to a rise in sea and air temperatures. The two species compete where they coexist, and *B. balanoides* has a higher growth rate and can therefore replace *C. stellatus*. However, *B. balanoides* can only breed successfully when temperatures are low, and feeds less efficiently at higher temperatures than *C. stellatus*. This allows *C. stellatus* to outcompete *B. balanoides* in warmer waters. A difference in the response of these species to an abiotic stress (water temperature) therefore underlies their relative competitive ability.

Environmental changes may also exert their effects via other types of biotic interactions. When populations of limpets are reduced by toxic effects of oil spills, the result is luxuriant growth of algal species that are normally grazed by limpets (Suchanek, 1993). Outbreaks of many fungal diseases in plants have been linked to unusually hot and moist conditions. Bacterial and fungal diseases of insects and vertebrates have been linked to changing climatic conditions. For instance, Ford (1982) outlined the case of botulism, which affects birds as well as humans and other mammals. This disease is caused by a bacterium which produces a neurotoxin. In shallow lakes, the toxin can build up under warm anaerobic conditions which allow the bacterium to thrive. The toxin can eventually reach levels where it causes massive mortality in waterfowl. Levels remain high in lakes until rainfall dilutes the toxin and increases oxygen levels. Severe outbreaks therefore tend to occur when conditions are hot and dry. Climate may also determine the distribution of diseases indirectly by affecting disease vectors. For instance, the distribution of tsetse flies in Africa is largely determined by climatic variables (Dobson & Carper, 1992). These flies are responsible for transmitting sleeping sickness (trypanosomiasis), a disease of humans and animals caused by a pathogen.

The effects of environmental changes on biotic interactions can be complex. This is illustrated by the effects of drought on the outbreak of insect pests. When plants are under drought stress, populations of insects feeding on plants may increase or decrease. Changes in insect populations can depend on the level of drought stress experienced by plants. For instance, in two-spotted spider mites (*Tetranychus urticae*) feeding on beans, English-Loeb (1990) found that the number of mites per plant decreased when beans experienced a low or moderate level of stress, but increased again when the drought stress was more severe (Figure 1.1). It has been suggested that plants become more nutritious for insects when they are under a severe stress.

Indirect effects of climatic changes can extend across different levels of a community of animals and plants. Effects on one species can have a cascade of effects through other levels. Lodge (1993) discussed the case of fish in temperate North American lakes. The fish species that occur together as assemblages in these lakes can be predicted by the size of the lake, its depth and the pH content of its water. Each assemblage of species is determined by the effects of these environmental factors on the predator at the top of the food chain. For instance, when lakes are covered by ice, oxygen levels in water fall. Large predatory fish such as bass and pike do not occur because they cannot survive these conditions. This enables the survival of species that are susceptible to predation, such as minnows and mudminnows. If conditions become warmer, pike and bass may become more common and force minnows out. Other examples of cascading effects triggered by environmental changes are given in the volume edited by Kareiva, Kingsolver & Huey (1993).

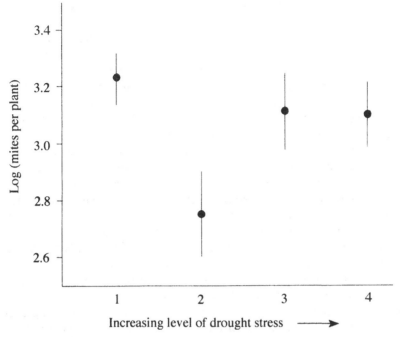

Figure 1.1. Mean number of two-spotted spider mites per bean plant subjected to four levels of drought stress. Means are averaged over five sampling times and two treatments involving the presence or absence of predatory mites. Error bars represent standard errors of the means. (Simplified from English-Loeb, 1990).

This brief overview indicates that organisms can be affected by environmental extremes in a variety of ways. Evolutionary responses to changing conditions can therefore be diverse. They may be related to direct effects, as in the case of increased survival under stressful conditions, or related to indirect effects mediated by biotic factors, as in the case of altered competitive ability and increased protection from predation.

Measuring the direct effects of environmental change

The detection of extreme conditions might seem relatively straightforward because physical features of the environment such as temperature, humidity and pH can be easily measured. However, if the impact of environmental changes on organisms is to be assessed, the conditions *actually* experienced by organisms need to be measured, along with the biological impact of these changes. As we will see, this is not an easy task, even when direct effects are being measured.

One difficulty in determining the conditions experienced by organisms is that the organisms themselves can modify their environments. The conditions experienced by plants and relatively sessile animals such as mussels and corals are fairly easy to determine, because these organisms have a limited ability to escape stressful conditions. In contrast, determining the environments of many animals can be difficult because of their ability to evade adverse conditions.

The extent to which animals can modify the conditions they experience can be illustrated by research on ectotherms. Many studies have demonstrated that lizards can maintain relatively constant body temperatures despite variation in the temperature of their environment. An example is the thermal response of the lizard, *Podacris hispanica*, an endangered species which inhabits islands in the Mediterranean (Castilla & Bauwens, 1991). Body temperatures of *P. hispanica* do not differ between cloudy and sunny days or show daily changes despite considerable variation in ambient temperature. The lizards are able to maintain body temperatures by basking in sunny positions at some times of the day, as well as by moving continuously between sunny and shady areas. The mean body temperature of *P. hispanica* in the field (34.0°C for females) corresponds closely with the temperature they select when placed in a gradient (34.7°C). However, this is higher than both the air temperature (23.6°C) and ground temperature (28.5°C) recorded in the field. Lizards are able to maintain an optimal body temperature by behavioural responses.

Another example of behavioural modification is provided by *Drosophila melanogaster*. This species lays its eggs mainly in rotting fruit. The conditions experienced by larvae and pupae will depend to some extent on where eggs are laid, and on larval movement around fruit and into nearby soil. The environmental temperature experienced by *Drosophila* in the field cannot be determined directly because of their small size. However, Jones *et al.* (1987) were able to indirectly assess temperature preference in *D. melanogaster* using a temperature-sensitive eye mutant. For this mutant, adult eye colour depends upon developmental temperatures experienced by young pupae. By collecting adults, Jones *et al.* (1987) were therefore able to determine temperatures experienced by pupae and compare them to air temperatures. Flies were collected at two locations in eastern U.S.A. At Beltsville (40 m elevation) the mean temperature experienced by pupae was 22.0±0.26°C and at Shenandoah (1000 m elevation) it was 20.7±0.19°C. While the temperature difference experienced by *Drosophila* at these sites (1.3°C) is considerable, it is much less than the 4.5°C difference in mean air temperature at the two sites. *D. melanogaster* were therefore able to modify the temperatures they experienced by behavioural means. Adult *Drosophila* can also modify their environments behaviourally. For instance, *Drosophila inornata* adults rest in microhabitats in Australian rainforests with low levels of physical stress (Parsons, 1992a). It can therefore be misleading to determine temperature stresses experienced by organisms from ambient conditions.

As well as defining the environments experienced by organisms, we need to assess the biological effects of the conditions organisms experience to determine if they are extreme. As mentioned in the previous section, these effects should preferably be measured in terms of fitness, or the success of organisms in leaving descendants. Unfortunately, it is very difficult to measure the fitness of organisms under field conditions. Fitness depends not only on the number of offspring an individual produces, but also on other factors such as the time offspring are born and the time they take to become sexually mature.

In many cases, it may only be possible to measure one aspect of fitness. For example, the effects of environmental changes on fish can be assessed by swimming ability or the number of eggs they produce. Environmental effects on insects can be measured by flight ability or the time adults take to reach reproductive maturity, while effects on plants can be measured by growth rate or the amount of seed plants produce. In these examples, effects are being deduced from changes in one or a few traits likely to contribute to the total fitness of an organism, and such traits are referred to as 'fitness components'. These components are often closely related to the life history characteristics mentioned in the previous section. Yet changes in one component may not necessarily reflect changes in overall fitness. This can be illustrated with a simple example. In many insects, an increase in temperature at which immature stages are cultured results in a smaller body size for the adults. This in turn leads to a decreased reproductive output because small adults produce fewer eggs than large adults, suggesting a decrease in fitness under high temperatures. However, by emerging at a smaller size and spending less time at immature stages, adults develop more quickly and become sexually mature at an earlier age. Increasing temperatures therefore result in a decrease in fitness when measured in terms of one trait, but an increase in fitness when measured in terms of a different trait.

Because of these difficulties as well as the amount of time involved in evaluating changes in fitness components under field conditions, other indicators of extreme conditions have been developed. The effects of extreme conditions are often evident from changes in the behaviour or appearance of organisms. Birds can lose feathers or fly erratically when stressed, while mammals can appear sluggish or develop dull coats. Crawling insects often move around in a disorientated fashion when they are stressed. When plants are under stress because of the absence of nutrients, leaves often change colour and may eventually be shed. However, these measures of stress are rather subjective, and a set of visible changes will often be useful only for a particular species.

More objective measures include the use of biochemical variables that can be assessed on any organism. A number of biochemical variables have been proposed for this purpose. These include measures of enzyme function, hormone changes, free amino acids and the concentration of energy carriers in cells. One widely-used general indicator of stress level is the adenylate energy charge (*AEC*). This is an index measuring the amount of metabolic energy available to cells for their metabolic processes. It is a ratio defined as

$$AEC = \frac{ATP + \frac{1}{2}ADP}{ATP + ADP + AMP}$$

and is based upon the high energy compounds adenosine triphosphate (*ATP*), adenosine diphosphate (*ADP*) and adenosine monophosphate (*AMP*). *ATP* is converted into *ADP* with the loss of energy, while a further loss of energy occurs when *ADP* is converted to *AMP*. The *AEC* ratio is therefore largest when there is abundant metabolic energy in the form of *ATP* in the cell. As organisms become stressed, the metabolic energy available to cells becomes reduced, and the *AEC* falls. *AEC*s decline in response to a wide range of stresses including salinity, temperature extremes, anoxia, desiccation and pollutants (Ivanovici & Wiebe, 1981). However, changes in *AEC*s are only indicative of extreme conditions when a drastic reduction in fitness is likely. This indicator and other indicators of extreme conditions do not necessarily show a linear relationship with fitness.

Another measurement of stress attracting increasing attention is the 'fluctuating asymmetry' (FA) of bilateral characters that can be measured on both sides of an organism (Zakharov, 1989; Parsons, 1990; Markow, 1995). Asymmetry is measured as the difference between the right and left sides of individuals. Normally, when a group of individuals has been measured for a particular trait, the mean difference for left measurements taken away from right measurements (R−L) is expected to be zero. Fluctuating asymmetry is usually taken to be the absolute value of this mean difference (|R−L|) and increases as individuals deviate from perfect symmetry.

FA has been studied for many paired characters in a variety of animals and to a lesser extent plants. Examples of paired characters include the numbers of scales of lizards, the number of bristles of flies, the length of veins on the wings of insects, and length measurements on the bones of mammals. It has often been suggested that FA provides a measure of how stable an organism's development has been, which in turn is expected to depend on levels of stress in an environment. As the level of any kind of stress during development increases, FA should increase, provided that the stress is sufficiently intense. Many experiments and observations support this conjecture, especially in relation to temperature extremes in *D. melanogaster* and other insects. The association between FA and stress is discussed further in the next chapter where we consider the effects that extreme conditions can have on the expression of traits and genes.

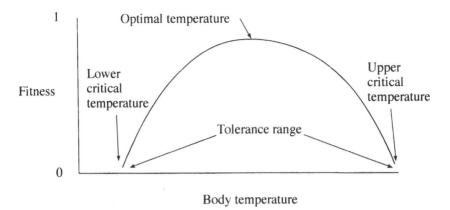

Figure 1.2. Two-tailed curve showing the effects of body temperature on the fitness of an ectotherm. (After Huey & Stevenson, 1979).

Environmental response curves

Although we are concerned with extreme conditions, we will also be considering how characteristics favoured at extremes fare under benign conditions. The usual way of portraying how the fitness of organisms with particular characteristics varies across a range of environments is to use response curves. In these curves, some measure of fitness is plotted against environmental conditions experienced by organisms.

Response curves will often be two-tailed. An example is provided by the typical response curve for an ectothermic animal plotted against body temperature in Figure 1.2 (Huey & Stevenson, 1979). Low environmental temperatures will result in low body temperatures and eventually a critical temperature will be reached (known as the 'lower critical temperature') where continued survival and reproduction of an organism is precluded. At the other extreme, an upper critical temperature can be defined in the same way, while maximum fitness occurs when animals are at intermediate temperatures. The fitness measure in such curves can involve a range of traits, such as an animal's reproductive output or its mobility as it searches for food or mates.

Two-tailed curves apply to many stresses. Dry conditions may be stressful to plants when leaves wilt and photosynthesis is reduced, while excessively wet conditions may be stressful because there is a decreased oxygen supply due to waterlogging. The salt content of water can be toxic to marine organisms at high concentrations and can cause osmotic stress at low concentrations. Animals need essential vitamins and minerals but these can become toxic when intake levels are too high.

One-tailed response curves form a better approximation for other stresses,

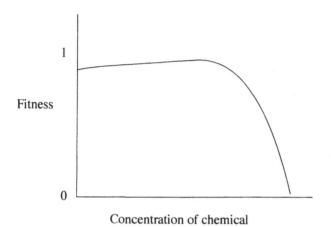

Figure 1.3. One-tailed curve showing the effects of changing concentrations of a chemical on the fitness of an organism. Note that low concentrations of chemicals often lead to a slight increase in fitness.

such as toxins which have few beneficial effects at low concentrations. Responses of animals to chemicals and heavy metals fall into this category, including responses to pesticides. A typical curve is given in Figure 1.3 which illustrates the increasingly toxic effects of a chemical. The response curve shows a slight increase in fitness at low concentrations of the chemical compared to situations where the chemical is absent. This is because chemicals that are usually toxic may be essential to the normal metabolism of organisms at low concentrations.

These two types of response curves should not be regarded as the only possibilities because curves can fall within these extremes. For instance, the availability of food for animals is often viewed as a one-tailed curve because a shortage of food is expected to lower fitness. However, an excess of resources can often decrease fitness, as in rodents whose longevity is increased when they have restricted access to food (Pieri *et al.*, 1992). Curves showing a small increase in fitness as food availability changes from being abundant to being restricted, followed by a sharp decrease in fitness as food shortages develop, are probably more appropriate for this type of stress.

Traits involved in stress responses

Organisms can deal with environmental extremes in a number of ways. Once an animal detects a stressful situation, several responses can take place. Animals may modify the environment they experience by moving away from a stress. This can involve a localized movement over a short distance. For instance, some insects evade pesticides by being repelled by the pesticide's odour and moving a few metres to an area where the pesticide is absent, while lizards can avoid heat stress by moving a short distance to a cool site in the shade. Even a small amount of movement may be sufficient to evade stressful conditions. Resting butterflies can evade a heat stress by changing the way their body is orientated towards the sun, minimizing the amount of wing area directly exposed to the sun's rays, while mussels can avoid desiccation stress when they are exposed during low tide by closing their shells. Behavioural changes unrelated to movement can also help to evade a stress. For instance, some animals evade desiccation stress by increasing their intake of water via the consumption of food with a high water content.

Behavioural modifications may be adequate for countering the effects of short-term environmental changes, but they become less effective as extreme conditions persist. When this happens, a more drastic response is required to evade a stress. One possibility is for animals to migrate to more favourable habitats. Many animals migrate long distances to evade seasonally stressful conditions. Another possibility is for animals to evade extreme conditions by entering an inactive phase. Many species can survive climatic extremes when they are in such phases. Examples include hibernation in mammals and diapause in insects. Entering an inactive or migratory phase normally requires major changes in an organism's physiological processes. These usually commence once conditions preceding the onset of climatic extremes are detected. For example, a common trigger for the development of inactive phases in an insect's life cycle is a reduction in daylength preceding the onset of winter.

If immediate behavioural responses are insufficient to evade a stress, a number of biochemical and physiological changes are usually initiated. The production of specific hormones may be triggered, which in turn can influence biochemical pathways that act to reduce the effects of a stress. Various changes in an organism's metabolism take place, including changes in the activity of enzymes and the synthesis of a special class of proteins known as the heat shock proteins.

If these metabolic changes successfully counter the effects of a stress, an organism can recover completely and be able to function normally in the new environment because of its increased level of resistance. Such a recovery is known as 'acclimation', and this process normally occurs before the onset of lethal levels of a stress. Animals often become acclimated to counter the effects of seasonal climatic changes, and increases in resistance can be spectacular. For instance, the upper lethal temperature of goldfish acclimated at 0°C is 27°C, but this can be increased to 41°C when goldfish are acclimated at high temperatures for a short period (Fry, 1958). This increase in survival often has a cost. For instance, when flies are acclimated at high temperatures, their survival of a heat stress is improved dramatically, but they also have a decreased ability to produce eggs (Krebs & Loeschcke, 1994).

These considerations indicate that traits involved in direct evolutionary responses can be complex. They may involve evading a stress or resisting its effects. Resistance in turn can be modified by acclimation, which depends on the way organisms perceive external conditions.

Traits involved in indirect responses can be even more difficult to measure because they involve interactions between organisms. Nevertheless some of the traits important in direct responses will also be important in responses to indirect effects. For instance, growth rate is likely to contribute to the competitive ability of an organism, because individuals with a rapid growth rate will outcompete slow growing competitors. Similarly, reproductive output will influence susceptibility to predation because individuals with a high output are more likely to produce at least some progeny that survive predation pressures. Many of these traits fall under the umbrella of 'life history' characters, those components of an organism's life cycle that together comprise the contribution the organism makes to the next generation.

Four evolutionary consequences of extreme environments

In focussing on the effects of severe environmental changes, we are considering only a subset of the areas that are normally studied by ecologists and evolutionary biologists. We believe that this emphasis is appropriate because of the mounting evidence that infrequent environmental changes superimposed on normal conditions are predominant in the evolutionary history of organisms. We start by considering four consequences of environmental changes that underlie the evolutionary importance of stressful conditions. These consequences appear prominently and repeatedly in later chapters.

(a) Environmental extremes influence the expression of phenotypic and genetic variation

The appearance of an individual is known as its 'phenotype', which is determined by the combined effects of an individual's genetic make-up (its genotype) and the environment it experiences. During development, genes act in combination with environmental factors to produce phenotypes. Phenotypic variation among individuals can range from being largely genetic in origin, to being largely determined by environmental variation. For instance, at an intermediate position in this range, variation in the body size of an animal species may be determined by a combination of genes controlling its growth rate and the food available in its environment.

Phenotypic differences among individuals are responsible for differences in fitness. Because natural selection depends on fitness differences, rates of evolutionary change driven by natural selection are determined by the amount of phenotypic variation present in a population. However, this variation has to be at least partly genetic if populations are to evolve. If phenotypic variation is completely environmental, any differences between individuals in one generation will not be passed to the next generation. For instance, consider the case of body size. In many animals, large males are often more successful at procuring mates than small males, resulting in a relatively higher fitness for large males, at least when the effects of size on other fitness components is ignored. If differences in body size do not have a genetic basis, then size will not change in a population despite such fitness differences. Rates of evolution in a population therefore depend on the amount of phenotypic variation and the extent to which it is genetic, a point that will be discussed in detail in Chapters 2 and 3.

Extreme conditions can influence both the expression of phenotypic variability and the extent to which it is genetically determined. One way this occurs is via the disruption of normal development. Under stressful conditions, the development of organisms is often disturbed, and this can lead to the appearance of characteristics not evident when organisms develop under normal conditions. For instance in *Drosophila melanogaster*, stress can produce abnormal wing venation patterns or abnormal body segments when individuals are exposed to chemicals or high temperatures early in their development. In this way, stresses can lead to the expression of an increased level of variability in many traits. Some researchers have argued that the production of novelties in stressful environments provides a source of variation on which natural selection can act. In other words, genes controlling the expression of novel characters can be selected once the environment is stressful, even though effects of these genes are not evident under optimal environmental conditions. Continued selection of the novel characters could favour genes that lead to the expression of novelties even in the absence of a stress. This has led to the proposal that major evolutionary shifts requiring novel characters will only occur when environments are stressful. In contrast, the development of organisms is considered too well buffered to allow novelties to arise when conditions are favourable.

There are also other ways in which environmental conditions influence the amount of phenotypic and genetic variation in a population. Perhaps the most direct is via mutation and the process of recombination. Stresses can increase both the rate at which new mutations arise, and the extent to which genes are shuffled during sexual reproduction by recombination. Stressful conditions can therefore help to generate new genes and new combinations of genes in a population, raising the possibility that the genome has evolved to respond in a flexible way to environmental conditions (Jablonka & Lamb, 1995). We examine these possibilities and the production of evolutionary novelties in the next chapter.

(b) Extreme conditions are periods of intense natural selection

Despite the widespread existence of stressful environments, these have not traditionally been given an important position in the process of evolution by natural selection. The reason for this appears to be partly historical. Darwin mainly regarded evolution as a gradual process involving intricate and complex interactions between organisms. He considered evolution to be driven by the process of natural selection operating largely through competition between individuals. Darwin (1859) considered that those individuals that could outcompete others were more likely to survive and reproduce, and were therefore more likely to pass on their characteristics to offspring. Characteristics that improved competitive ability are thereby selected, and these increase in frequency in a population. The population evolves as traits improving competitive ability are favoured, and evolutionary change is largely driven by competitive interactions between individuals.

In contrast, Darwin assigned a much smaller role to natural selection as a consequence of environmental changes. He acknowledged the existence of extreme climatic conditions as indicated by the following quotes from 'The Origin of Species' (Darwin, 1859):

> Climate plays an important role in determining the average numbers of a species, and periodical seasons of extreme cold or drought seem to be the most effective of all checks

> When we reach the Arctic region, or snow-capped summits, or absolute deserts, the struggle for life is almost exclusively with the elements

He accepted that natural selection could occur when individuals differed in their ability to cope with stressful conditions, and that traits related to stress responses rather than competitive ability could be selected under these extreme conditions. However, such references are scattered in Darwin's work which is more concerned with the role of competition in generating evolutionary change. Darwin did not really incorporate extreme situations into his major arguments. Similarly, most of Darwin's successors did not emphasize the effect of environmental stress on evolutionary change. A notable exception was the Russian biologist, I. Schmalhausen, perhaps because he lived in adverse climatic conditions, and so appreciated their evolutionary consequences!

There is little doubt that periods of environmental change have the potential to cause intense natural selection, leading to rapid evolutionary change via the process initially outlined by Darwin. We can illustrate the process by referring to a selection experiment of the type carried out in agriculture where animal and plant breeders are attempting to increase growth rate, production, yield or some other economically important trait. Consider a population where the yield produced by individual plants is spread around a mean value.

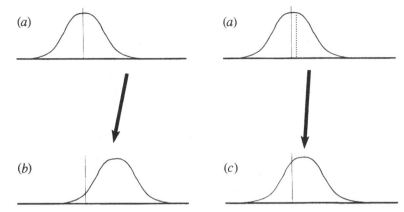

Figure 1.4. Changes in the distribution of a trait when a population is exposed to directional selection. Individuals to the right of the dotted line in (*a*) are selected as founders for the next generation. When selection is intense and only those individuals with extreme values of a trait are favoured (*b*), there is a large shift in the distribution of a trait in the next generation, whereas the shift can be small when selection is weaker (*c*). A shift in the distribution of a trait will only occur if some of the variation among individuals is genetic and therefore passed on between generations.

Plant breeders are likely to be interested in selecting those individuals with a higher yield than this mean value, as illustrated in Figure 1.4. If the variation in yield between individuals is partly determined by their genes (i.e., the variation has some 'genetic component'), then offspring of parents with a yield higher than the mean value will also tend to have yields higher than this value. In other words, the mean yield of a population consisting of offspring has become higher than the mean yield of the parental population. Yield has therefore increased over this interval of one generation because genes with a high yield have been favoured at the expense of those with a low yield.

The extent to which the yield of the offspring has shifted away from the mean value of the parental population can be predicted from the theory of quantitative genetics. Basically, as well as depending on the degree to which a trait is genetically determined, the response also depends on what is known as the 'intensity of selection'. This intensity is determined by those parents that are selected to produce the offspring generation. If only parents with yields far greater than the mean are selected, then the selection intensity will be high (Figure 1.4(*b*)). In contrast, the selection intensity will be much lower if all parents higher than the mean value are selected to produce offspring (Figure 1.4(*c*)). The higher the selection intensity, the greater the difference between mean yield of the offspring generation and that of the parental generation, and hence the faster the response to selection.

It should be emphasized that such a response depends on a proportion of the variation in yield being controlled by genes, since variation among individuals is also influenced by the environment. For instance, yield will differ depending on whether a plant is grown on rich soil compared to poor soil. The greater the extent that genes rather than the environment contribute to the variation, the greater the similarity between parents and their offspring and the greater the response to selection. We return to this topic in Chapter 3 when we consider examples of evolutionary changes in populations.

Because selection can be intense at times of severe environmental stress, evolutionary change can be rapid. When a population encounters an environmental stress, only those individuals with characteristics that enable them to withstand the effects of the stress will survive. If the mortality in a population following stress is 95%, the surviving 5% are likely to possess characteristics that deviate considerably from the mean of the population before selection. In contrast, traits involved in responses to an environmental stress may not change much for many generations in the absence of this stress. Traits that affect survival and reproduction under stressful conditions can therefore only evolve rapidly when an environment changes.

We can illustrate this process with an example. In the Galápagos Islands visited by Charles Darwin on the voyage of the *Beagle*, there is a remarkably wide range of habitat-specific finch species that differ in morphology, particularly in the size and shape of their bills. One of these species of Darwin's finches (*Geospiza fortis*) has been the subject of a long-term investigation into natural selection by P. R. Grant and his associates (Grant, 1986). The birds on an island were ringed and their body dimensions were measured. This allowed researchers to follow selection on body morphology by repeatedly measuring the birds and their progeny. A drought in 1977 led to a drastic reduction in the small seeds that formed the main food supply for *G. fortis*. This in turn resulted in starvation and an 85% reduction in the size of the finch population. The period of intense selection led to a change in the bill and body sizes of the population (Figure 1.5). Before selection, the mean weight of the birds was 15.8 g. This increased to 16.9 g after selection. Large birds were favoured by natural selection, because they had large bills and were able to crack the large and hard seeds that formed the remaining food source during the drought. Birds with smaller bills were unable to deal with the large seeds. This morphological change persisted in the next generation because variation in body dimensions in *G. fortis* is largely controlled by genes, with only a small contribution from the environment. Stress therefore led to a rapid evolutionary change in body morphology. In contrast, only slight changes in body size were recorded in years when there was no severe drought. More examples of such rapid changes in populations are described in Chapter 3.

It is possible that many of the evolutionary changes seen in populations stem from such periods of selection even if these periods occur infrequently. The fact

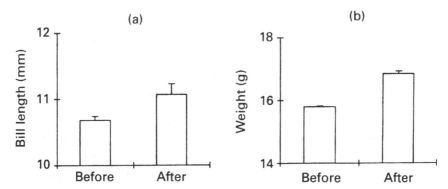

Figure 1.5. Changes in (a) bill length and (b) weight in a population of Darwin's finches (*Geospiza fortis*) following a drought. (Plotted from data in Grant, 1986).

that adaptive changes in many traits involve a fairly simple genetic basis provides suggestive evidence for selection under extremes, because many genes might otherwise be expected to contribute to a selection response as discussed in Chapter 3. Moreover, selection on traits that are normally invariant may only be effective if conditions are extreme enough to generate phenotypic variability in such traits. There is also evidence that extreme conditions enhance fitness differences among genotypes, exposing them to natural selection.

Rapid changes under extremes can alter traits other than those under selection. This results from the phenomenon of 'pleiotropy', where the same gene influences more than one trait. A gene affecting body colour could also influence traits such as sensitivity to ultraviolet radiation, camouflage, or rates of heat absorption. Genes influencing levels of a hormone are expected to have pleiotropic effects because hormones tend to control multiple physiological processes. In plants, genes that affect root growth will influence not only water uptake, but also nutrient uptake and the stability of plants in windy conditions. Pleiotropy is likely to be the rule rather than the exception, and genes influencing stress responses will also influence other traits. Some of these traits may affect fitness under non-stressful conditions, a point we return to in Chapter 4 where we discuss how interactions among traits can restrict the extent to which populations respond to selection.

(c) Selection under extremes can produce characteristics that are not normally favoured

When organisms encounter extreme conditions, they may survive because of a life history pattern that enables them to evade stressful conditions. Evasion can occur because animals move away, adopt a stress-resistant life-cycle stage, or reduce the amount of effort they expend on activities such as reproduction in order to increase the resources they have available for survival. Organisms may also survive because they are inherently resistant to stressful conditions. For instance, plants with a slow growth rate can persist in environments where resources are sporadically limiting.

Theoretical work has been used to predict the optimal life history that organisms are expected to show under different environments, utilizing these types of evasion characteristics. For instance, theoretical predictions have been made about the number of dormant seeds that a plant should produce, in relation to its likelihood of encountering adverse conditions. Predictions are tested against empirical data to see if an organism's life history can be related to the environmental extremes it experiences.

In general, these predictions show that occasional extreme events can have a large influence on the life histories of organisms. This is because individuals with a particularly high fitness may comprise those that do well under poor conditions. We can illustrate this by referring to the simplest case. When individuals live in a constant environment, where generations are non-overlapping, the optimal type of individual is the one with the highest score for fitness. However, once environmental variability is introduced, the estimation of fitness becomes more complex. Intuitively, we might compare the average fitness of different types of individuals across a range of environments. So if the fitness of one type in a range of N environments is represented by the symbols X_1, X_2, X_3 and so forth to X_N, its average fitness is given by

$$\overline{w} = \frac{X_1 + X_2 + X_3 \cdots\cdots X_N}{N}.$$

However, when conditions fluctuate, it can be shown that the average fitness of a genotype is given by its geometric mean, rather than its arithmetic mean (Cohen, 1966; Gillespie, 1973), ie.

$$\overline{w} = \sqrt[N]{X_1 X_2 X_3 \cdots\cdots X_N}.$$

This means that fitness in a bad environment can contribute disproportionately to the overall fitness of a particular type. Consider the case where the relative fitness of one type (A) is 0.8 in a good environment, and 1.0 in a bad environment. Contrast this with another type (B) having a fitness of 1.0 in a good environment but only 0.4 in the bad environment. If there are four good years followed by a single bad year, the arithmetic average fitness of A is 0.88, and that of B is 0.84. In contrast, the geometric mean fitness of A is 0.83, compared to 0.87 for B. In other words, the overall relative fitness of these two genotypes is influenced by the low fitness of A in a poor year even though it does better in most years, and B will be favoured overall. This highlights that the optimal type likely to evolve in a population can be difficult to predict unless there is information on the relative fitness of types over several environments including extreme conditions.

We can relate these fitness differences to a hypothetical example. Consider an insect species consisting of two types of individuals, one of which produces diapause eggs. Under continuously favourable conditions, the type that does not produce the diapause eggs will be at an advantage, because its progeny will hatch immediately and produce their own progeny. However, this type may have a very low fitness when conditions deteriorate and almost all insects die unless they are in a diapause stage. In contrast, the type producing diapause eggs will have a high fitness under these conditions because its eggs will hatch when conditions become favourable again. Even when severe winters are rare, the overall fitness of the diapausing type will be relatively higher because rare conditions have a large effect on the geometric mean. We consider the effects of environmental extremes on the types of life history patterns that organisms have evolved in Chapter 5.

Over a long timespan, exceedingly rare periods of intense stress may still influence characteristics possessed by organisms. The fossil record contains evidence of several periods of mass extinctions coinciding with environmental changes on a global scale. Some groups of organisms have been more successful at surviving these periods than others. Surviving groups tend to have low metabolic requirements or mechanisms to evade stressful periods (Chapter 6).

(d) Extreme conditions can influence population size and result in extinctions

Periods of severe environmental stress will reduce the size and age structure of populations as mortality occurs and the reproductive output of organisms decreases. In extreme cases, stresses may result in the extinction of populations and a reduction in the area occupied by a species. These changes can have evolutionary consequences by influencing the ability of populations to adapt to the stressful conditions and to other environmental changes.

When a population remains at a small size for several generations, the degree of inbreeding in a population is expected to increase. Because members of small populations will often be related, individuals will tend to mate with related individuals, leading to inbreeding. Most populations contain many rare recessive genes which can cause a drastic reduction in fitness when they become homozygous. The effects of these genes are seldom evident in large populations because they are hidden by their recessive nature. Examples in human populations are the many genes causing genetic diseases such as phenylketonuria and alkaptonuria. When inbreeding occurs, such genes become expressed more frequently because matings between related individuals will often produce offspring that are homozygous for these genes. This causes a reduction in the mean fitness of individuals in a population. In extreme cases, inbreeding may lead to the extinction of a population. Inbreeding is particularly important in the conservation of captive populations that are held at a small size. However, inbreeding may be less important in natural populations because these often rapidly increase in size after short stressful periods. We discuss this topic in Chapter 7 where we consider the conservation of genetic variation in populations to enable them to counter the effects of future environmental changes.

As well as being more prone to inbreeding, small populations are also more prone to losing genetic variation by the process of 'genetic drift'. Populations are normally so variable at the genetic level that no two individuals are genetically the same. The process of mutation continually adds new genetic variants to a population, and these variants are shuffled into new combinations by sexual reproduction. Sex enables genes from different individuals to come together and form new combinations in their offspring. However, there is always a possibility of a gene being lost from a population. If a gene is rare, it will be carried by only a few individuals. It is possible that, by chance, none of these carriers reproduce in a particular generation. Perhaps the individuals are sterile, or their progeny die accidentally. In this case, the rare gene will be lost. It may not reappear in a population until a migrant brings the gene from a neighbouring population, or until it arises anew by mutation.

The loss of rare genes is much more likely in small populations than in large ones. This should be fairly obvious. If a gene occurs at a frequency of only a few per cent, it is more likely to be present in a population of 100 individuals than in a population of only 10 individuals. As a consequence, populations can become genetically less variable over time if their size remains low for many generations. Rare genes can also be lost when there is a short-term decrease in population size. This has obvious implications when the missing genes suddenly become important for survival. For example, a new disease might arise in a population and kill all those individuals without a rare resistance gene. Extinction is more likely in such situations if a population's size is small. Size can therefore determine the potential for organisms to adapt to environmental changes. Maintaining high levels of genetic variation is of particular importance in the conservation of species to ensure their long-term survival (Chapter 7).

A final consequence of drastic changes in population size is that many of the environments previously occupied by a species can become vacated by it, creating empty 'ecological space'. This space may subsequently be colonized by new populations of the same species or of a different species. It has been proposed that this process can facilitate evolutionary changes, since empty space may become occupied by new evolutionary novelties. These can then spread and undergo further evolution. This process is considered by many researchers to be important in macroevolutionary changes detected in the fossil record, and it is considered in Chapter 6 where we discuss how changes in the fossil record can be linked to periods of environmental stress.

Overviewing implications of the four evolutionary consequences

The above consequences of environmental changes form the basis for our discussions about the role of these changes in evolution. In Chapter 4, we examine factors that limit evolutionary responses to environmental changes. One of the issues we discuss is why species often have restricted distributions. Distribution limits indicate that species cannot exist under environmental conditions outside their range. Species appear to be limited in their ability to adapt to environmental stresses occurring at the edge of their ranges. We examine the reasons for such limits. Closely related to this area is the question of 'tradeoffs' between traits. Tradeoffs arise when selection for one phenotype is somehow opposed by selection for another phenotype. This can restrict evolutionary responses when both phenotypes are favoured by selection. The intense selection associated with environmental changes and the effects of stress on population size are key issues in understanding limits to adaptation.

We also consider evolutionary changes in the fossil record (Chapter 6). We examine the origin of large phenotypic changes studied by palaeontologists. Evolutionary change can be studied over a much longer timespan than in populations existing today. Palaeontologists disagree about the relative importance of competition and environmental changes in causing evolution in the fossil record. We consider this controversy and the role of extinctions in triggering periods of evolutionary divergence. The effects of environmental changes on population size and the effects of stress on phenotypic variation are often seen as key elements promoting the development of new evolutionary lineages in the fossil record.

Finally, we discuss the issue of conservation (Chapter 7). Because of the increased environmental pressures arising from human activities, rapid environmental changes are predicted in the next century. Populations need to adapt to these changes if they are to avoid extinction. We consider the likelihood of adaptation to changes including the predicted increase in global temperatures. We also examine steps that might be taken to maintain variation in genes important in adapting to these types of responses. Information about the likely effects of environmental stresses on population size needs to be considered in management programmes for threatened species. In addition, the possibility of tradeoffs between traits favoured under stressful and benign conditions can have important implications in the conservation of genes that are useful in countering stressful conditions.

CHAPTER 2

Variation under extreme environments

In the traditional view of evolution by natural selection, individuals possessing a particular inherited characteristic (say X) arise by chance. Assume that X results in increased fitness in a new environment because individuals leave more progeny. X will be passed on to progeny, and as a consequence it will increase in frequency in a population as adaptation occurs to the new environment. Under this view of evolution, the environment acts solely as a selective agent on the inherited variation. The environment does not influence the extent to which a characteristic is inherited, or the likelihood that a new characteristic is produced.

However, environmental changes can directly affect the genetic information carried by organisms and the expression of variation in characters. Environmental effects have been detected by looking at the way different conditions influence mutation rates, recombination rates, the stability of an organism's development, and the way genes interact with the environment to produce phenotypes. We will show how these effects influence variation at the molecular and phenotypic levels, and how they can be substantial under stressful conditions.

We also consider whether the effects of extreme conditions are important in the evolution of traits that are generally invariant. Laboratory studies suggest that variation in these traits is markedly increased under stress, and invariant traits may not evolve unless stressful conditions are present. Evolutionary changes in such traits can involve a fairly simple genetic basis.

Mutations: random or directed by environmental challenges?

Evolution depends on inherited phenotypic variation in populations. There are two ways new variation is generated. The first is mutation, the process whereby new alleles arise because of changes in the genetic code. The second is recombination, the process that generates genetic variation in progeny when crossing-over occurs in chromosomes of the parents. This results in the production of new combinations of genes in the progeny generation.

Adverse environmental conditions can increase mutation rates. We are all familiar with the effects of ionising radiation and toxic chemicals on mutation rates. These represent examples of mutations increasing under stressful conditions, albeit artificial conditions often imposed by humans.

Mutation rates can increase under other types of stresses, including those encountered by organisms in nature. This was noted over 40 years ago by Schmalhausen (1949), who cited increases in mutation rates from factors such as extreme temperatures, intensive insolation, extreme limits of humidity and chemical influences of salts. Effects of high temperatures on mutation rates have now been demonstrated in a wide array of taxa, including *Drosophila*, microorganisms, fungi, and plants. Cold shock has also been shown to increase mutation in *Drosophila melanogaster*, suggesting that temperature extremes are important rather than temperature *per se*.

Some increases in mutation rates may be associated with mobile genetic elements known as 'transposons', which insert themselves into the DNA of host organisms. These elements can cause mutations when they insert into a gene. After insertion, the DNA sequence of the gene is altered, preventing it from coding for a functional gene product. The gene can therefore become inactivated or produce a modified product. It has long been known that environmental conditions can elevate mutation rates in maize plants via transposons (McClintock, 1984). There is also evidence that transposition rates increase with temperature stress in yeast and in *Drosophila*. For instance, Ratner *et al.* (1992) described the induction of transposition in a mobile element following heat shock in *D. melanogaster*. Transposition rates were in the order of 10^{-2} following a heat shock, compared to $<10^{-4}$ in controls, and transposons were more likely to insert in some parts of the genome than others.

In some plants, genetic changes that arise much more commonly than normal mutations can be triggered by changes in environmental conditions (Cullis, 1987). The most widely studied effects are in flax where heritable changes in weight and height have been found after plants are grown for a generation in controlled environments with imbalances in nutrients or with specific temperature treatments. At the molecular level, these environments induced changes in a type of DNA known as 'highly repetitive' DNA. Part of an organism's genetic make-up consists of genes that occur in multiple copies which comprise this highly repetitive class. On the other hand, other genes occur only as a few copies or as single copies. Highly repetitive DNA sequences include those coding for subunits of ribosomes. These genes are known as the 25S, 18S and 5S genes. When flax is cultured in stressful environments that trigger changes in weight and height, the number of copies of these genes is increased. However, it is unclear how changes in ribosomal genes cause changes in size and weight.

It is possible that mechanisms increasing mutation rates under stress are adaptive as argued by Jablonka & Lamb (1995). An increase in mutation rates may allow organisms to generate genetically variable offspring, especially under adverse conditions when populations can face extinction. Most new mutations will tend to have a low fitness, and many are unlikely to survive (Fisher, 1930). However, a new mutation may occasionally allow organisms to cope with the adverse conditions. On the other hand, it is also possible that changes in mutation rates with stress are not adaptive. An increase in mutation under stressful conditions could be a consequence of organisms being in a stressed state.

If mutation rates are under selection, populations should exhibit optimum mutation rates. If mutation rates are high, organisms may produce many progeny with deleterious effects normally associated with new mutations. However, if mutation rates are low, organisms may rarely produce progeny with beneficial new mutations that enable adaptation to adverse conditions. Natural selection therefore should adjust mutation rates to intermediate levels. By exposing a population of *D. melanogaster* to irradiation for more than 600 generations, Nöthel (1987) demonstrated that mutation rates may be selected in this manner. Because irradiation increases mutation rates, they are expected to decrease under irradiation if an optimal level of mutation exists in populations. In agreement with this prediction, genetic factors evolved in the irradiated population that reduced the frequency of mutations induced by irradiation. In addition, the mutation rate of the selected population was lower than expected in the absence of irradiation. This rate increased again when the population was maintained without irradiation for many generations. These data therefore suggest that mutation rates can evolve in response to environmental conditions. However, it has not been shown that an increase in mutation rate *specifically* induced by stressful conditions is adaptive, rather than a consequence of the conditions themselves.

While it is generally accepted that extreme conditions increase mutation rates, we normally assume that this increase randomly affects genes. Mutations are therefore as likely to occur at loci under selection as at those not under selection. Recently, however, the random nature of this mutational increase has been challenged, with the proposal that stress increases mutation rates specifically at loci under selection.

Most of the evidence arguing for a non-random mutation process has come from bacteria. In *E. coli*, under conditions of prolonged and intense stress, such mutations have been postulated for several loci. An example is provided by Hall's (1990) experiments on mutations associated with the biosynthetic pathway for the amino acid tryptophan. Mutations were used that rendered bacteria unable to synthesize tryptophan. These mutations were of a particular type known as 'base substitutions', involving the substitution of a base in the DNA strand by another base. Mutant strains could not grow in the absence of tryptophan, but entered a stationary phase. Hall could therefore test for the frequency whereby mutant strains underwent an additional mutation event to allow them to synthesize tryptophan once again by following the growth of bacterial colonies. Bacterial cells were placed on medium with a small amount of tryptophan to enable small colonies to be established from each cell before the tryptophan was exhausted. The starved colonies were then examined for growth to see if mutations allowing further growth occurred.

Hall found that many of the colonies showed growth after a few days, at a frequency much higher than expected from the normal mutation rate. Because there was no detectable cell division during the starvation period, these mutations could not have arisen due to the continued production of new cells. Instead, the appearance of these mutations depended on the time bacteria spent in a starved state. Hall also showed that an increase in mutation rate was not detectable at other loci that were not under selection. This led to the conclusion that the increase in the mutation rate was aimed at the specific loci under selection, rather than simply involving an overall increase in the mutation rate of all loci. The process that produced these mutations was so powerful that it generated double mutations at rates approaching the rates of single mutations under identical conditions. For example, if two mutants are required for growth and these normally occur at a frequency of 10^{-5} each, then the expected occurrence of mutations in both genes is 10^{-10}. However, under directed mutation these occurred at a frequency of 10^{-5}, an increase of 100 000 times.

Unfortunately, interpretations of such experiments have been criticized for three reasons. First, the experiments often lack adequate controls (Lenski & Mittler, 1993). For instance, in Hall's experiments, starvation may have had a non-specific effect on the mutation rate. There are different mechanisms whereby non-specific mutation rates can be altered. As a consequence, scoring mutation rates at one other locus may not represent an adequate control. Mutations at the other locus acting as a control may occur via a different mechanism that is not influenced by starvation stress. A second problem with these experiments is that the mutation rate is defined in terms of the number of mutations occurring per generation. When cells are not dividing, such as when they are in a starved state, no new mutations are expected to occur. However, MacPhee (1993; MacPhee & Ambrose, 1996) has pointed out that the DNA of cells in a stationary phase can still be damaged. When this damage is repaired, mutations may occasionally be produced because there are bacterial repair pathways that produce errors, resulting in 'high' mutation rates per generation. The likelihood of DNA being damaged or repaired by an error prone pathway is not uniform at all points along the bacterial chromosome, producing different mutation rates for different genes. Finally, any hypothesis about mutations being directed is unlikely to become accepted unless a plausible mechanism is found. It is presently hard to envisage how such a mechanism might operate.

Even if bacterial mutations were at least partially directed, their role in the adaptive evolution of microorganisms and multicellular organisms remains unknown. From an evolutionary point of view, any process increasing the specificity of mutations could be important since the supply of variability is increased, as is the probability of adaptation to specific stresses. Conversely, the cost of producing deleterious mutations at other loci would be decreased.

Recombination and stress

Recombination requires crossing-over between homologous chromosomes, which is detected by following what happens to a pair of genes located on the same chromosome. Consider a cross between two genetically different strains. If one parental strain has the A_1 and B_1 alleles at two loci, and the other parental strain only has the A_2 and B_2 alleles, then a cross between these strains will produce an F1 individual which is heterozygous at both loci. Crossing-over between the two loci in the F1 can then be detected by the production of genotypes containing recombinant gametes (A_1B_2 and A_2B_1).

It has been known for many years that levels of crossing-over depend on the environment. Plough (1917) examined recombination in a central region of the second chromosome of *Drosophila melanogaster*. He showed that recombination increased when the culture temperature was raised or lowered from the non-stressful temperature of 25°C. At 13°C and 31°C, which are close to extremes where populations become extinct under laboratory conditions, there was a threefold increase in recombination giving a U-shaped curve (Figure 2.1). In the field, it is difficult to collect flies under such conditions, confirming that these temperatures are marginal for *D. melanogaster*.

Many subsequent experiments with *Drosophila* have confirmed that recombination increases at temperatures above and below those normally used to culture these insects. Studies in a number of other eukaryotic organisms including the nematode, *Caenorhabditis elegans*, tomatoes, and several fungi give parallel results (Parsons, 1988; Jablonka & Lamb, 1995), particularly when experiments include the entire temperature range over which fertile offspring can be produced. Stresses other than extreme temperatures have also been shown to increase recombination, such as starvation (in *Drosophila*) and crowding (in mice). Increases in recombination may therefore reflect a response to stress in general.

It is not known how stress increases recombination rates. In mice, Borodin (1987) found that immobilisation, heat, or social interactions between males inhibited the synthesis of DNA in testes. It was suggested that hormones released in response to stresses such as these inhibit the repair and replication of DNA, and that this may in turn disrupt pairing between chromosomes and influence crossing-over which occurs after pairing. However, more work is needed to elucidate the actual mechanism that is involved.

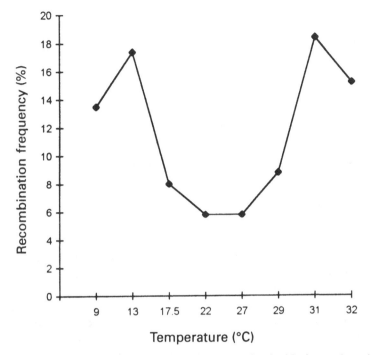

Figure 2.1. Recombination frequency for the black-purple region of chromosome 2 of *Drosophila melanogaster* plotted to show the effect of temperature on recombination. (After Plough, 1917).

From an evolutionary perspective, we need to consider whether an increase in recombination frequency under adverse conditions is adaptive, as in the case of mutation. As mentioned earlier, recombination generates new combinations of alleles. It is possible that some of these combinations may be better adapted to adverse conditions encountered by progeny than the allele combinations of parents. Organisms could therefore be at an advantage if they increase their recombination rate under stressful conditions, increasing the likelihood that they produce progeny that are better adapted than themselves. While this explanation is plausible, it is also possible that the increase in recombination rate is simply a consequence of an organism living under stress. Under these conditions, organisms may have fewer resources to devote to the synthesis and repair of their DNA. An increase in recombination could therefore be a consequence of having to allocate resources to other tasks. This explanation implies that the increase in recombination is not adaptive.

To distinguish between these possibilities, we can ask if adaptive changes in recombination frequency occur in response to a changing environment as in the case of mutation. If recombination frequencies are adaptive, they should increase when organisms are exposed to adverse environmental conditions

Table 2.1. *Changes in recombination frequency in chromosome 2 of*
D. melanogaster *following exposure to constant or variable temperatures for*
26 generations

Standard errors are given in brackets.

Population exposed to:	Recombination frequency (%)		
	b–cn	cn–vg	b–vg
Constant temperature	10.14 (2.36)	7.30 (1.45)	13.79 (3.21)
Variable temperature	14.30 (1.85)	12.95 (1.64)	24.97 (2.37)

Source: After Zhuchenko *et al.*, 1985.

that select for new combinations of alleles. No such increase is expected when populations are exposed to conditions that are optimal and constant. There is evidence that recombination rates evolve when populations are exposed to different environmental conditions. Zhuchenko, Korol & Kovtyukh (1985) exposed populations of *Drosophila melanogaster* either to temperatures that fluctuated daily between 15°C and 32°C or to a constant temperature of 25°C. Recombination was scored between two pairs of genes after populations had been held under these conditions for 26 generations. Recombination frequencies were substantially higher in the population exposed to the fluctuating conditions than in the one exposed to the constant environment (Table 2.1). This suggests that a high recombination rate is favoured when populations are exposed to conditions fluctuating between opposing extremes. In addition, experiments have shown that recombination frequencies can increase when populations are exposed to directional selection pressures for traits such as changes in behaviour or resistance to pesticides (reviewed in Korol & Illiadi, 1994). When increased resistance is favoured for many generations, organisms with relatively higher recombination rates in some parts of their genome are at a selective advantage. This may be a general trend when organisms adapt to new environmental conditions.

These types of experiments indicate that rates of recombination can increase when populations are exposed to stressful conditions that impose selection in one direction or changing directions. However, they have only considered changes in recombination rates irrespective of environmental conditions. It remains to be seen if adverse conditions can select for an increase in recombination rate that occurs *specifically* under stress and not when organisms are exposed to favourable conditions.

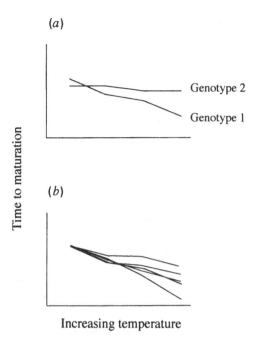

(*a*)

Genotype 2

Genotype 1

Time to maturation

(*b*)

Increasing temperature

Figure 2.2. Reaction norms of genotypes for the time taken to mature in response to increasing temperature. In (*a*), genotype 1 is plastic while genotype 2 shows low plasticity. In (*b*), the degree of canalization decreases as temperature increases. (Simplified from Stearns & Kawecki, 1994).

Reaction norms, plasticity and canalization

Before starting the next section, we briefly define terms that are used to describe the way the expression of the genotype and phenotype change with environmental conditions. The phenotypes produced by a genotype across a range of environmental conditions is known as its 'reaction norm'. For instance, in Figure 2.2(*a*) we plot the reaction norm of two genotypes that influence the age at which an organism reaches sexual maturity. In the case of genotype 1, time to maturity decreases with environmental temperature, whereas genotype 2 shows the same maturation time regardless of temperature. These genotypes therefore have different reaction norms.

The other term for describing the way phenotypic expression changes with the environment is 'plasticity'. In Figure 2.2(*a*), genotype 1 is plastic in the sense that its phenotypic expression depends on the environment. In contrast, genotype 2 is not plastic because it is relatively invariant.

Reaction norms determine the extent to which a trait is 'canalized'. Consider the reaction norms for a range of genotypes in Figure 2.2(*b*). At low temperatures, all genotypes have a similar age at maturity; in other words, the same phenotype is produced regardless of genotype. However, at high temperatures, the genotypes produce different phenotypes. We use canalization to describe the phenotypic variation produced by different genotypes. Because the same phenotype is produced at low temperatures, age at maturity is highly canalized under these conditions. In contrast, the degree of canalization is less at high temperatures. As discussed in Stearns & Kawecki (1993), a high degree of canalization does not imply a low level of plasticity, because phenotypes produced by genotypes can still change with environmental conditions. For instance, in Figure 2.2(*b*) the level of plasticity is high, but the trait is highly canalized.

Despite this, many traits have both a high degree of canalization and a low level of plasticity. For instance, in *Drosophila*, the number of bristles on the scutellum, a discrete region on the upper part of the thorax, is almost always four. Such characters that are almost invariant are useful for taxonomic work because they can define species and genera. Genitalia are particularly useful in this regard, because these cannot vary much in order for males to mate successfully with females from the same species even when they are reared under different conditions. Traits with a high degree of canalization and low plasticity are buffered against accidents or environmental changes that occur during the development of an organism, because any deviations from the norm can result in a drastic loss of fitness.

Phenotypic variation under extreme conditions

As well as increasing recombination and mutation rates, adverse conditions can increase the frequency of abnormal phenotypes in a population, thereby decreasing the extent to which traits are canalized during development. For instance, Zakharov (1989) investigated morphological variability in the lizard, *Lacerta agilis*, when embryos developed at a range of constant temperatures (20, 25, 30 and 35°C). Variability was scored by the arrangement of scales on the lizard's head, which can deviate from the normal pattern in a number of different ways (Figure 2.3). Zakharov found that deviations occurred less frequently when lizards developed at the intermediate temperature of 25°C than at the extreme temperatures. The highest incidence of abnormalities occurred at 20 and 35°C. Similar results have been obtained in experiments on the development of vertebrae in the snake, *Natrix fasciatus* (Osgood, 1978). The frequency of abnormalities in vertebrae is increased when embryos were exposed to both high and low temperatures. Gravid female snakes prefer a body temperature around 26.5°C, and this corresponds closely with the temperature range (25–27°C) where abnormalities in embryos are at a minimum. In both snakes and lizards, abnormal phenotypes therefore

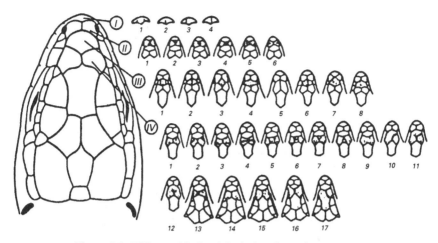

Figure 2.3. Different kinds of deviation from the usual scale patterns (I–IV) in the sand lizard *Lacerta agilis*. (From Zakharov, 1989).

occur at a higher frequency under extreme conditions compared to optimal conditions, and the degree of canalization is reduced. There is also evidence that stressful conditions trigger abnormal phenotypes in plants (Mabberley & Hay, 1994).

To look at the effects of environmental conditions on phenotypic variability more generally, we need a different way of considering variability across a range of organisms. In this context, one measure attracting increasing attention is the fluctuating asymmetry (FA) of bilateral characters that can be measured on both sides of a body (Zakharov, 1989; Parsons, 1990) and was mentioned in the previous chapter. Asymmetry in characters is measured by comparing differences between the right and left sides of the body. Normally, when a group of individuals has been measured, the mean difference for left measurements taken away from right measurements (R–L) is expected to be zero. Fluctuating asymmetry is usually taken to be the absolute value of this mean difference.

FA has been studied for many paired characters in a variety of animals. Much of this literature has been reviewed by Zakharov (1989), Parsons (1990) and Markow (1995). Examples of paired characters include the number of scales of lizards, the number of bristles of flies, the length of veins on the wings of insects, and length measurements on the bones of mammals. It has often been suggested that FA provides a measure of the stability of an organism's development. As the level of stress of any kind during development increases, FA and its variability should increase, provided that the stress is sufficiently intense to influence developmental processes. If this is true, FA should correlate with other measures of morphological variability such as those

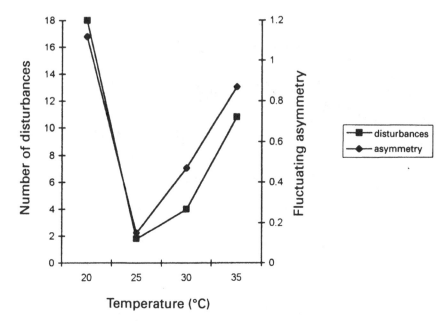

Figure 2.4. Effect of egg incubation temperature on the production of morphological abnormalities in the sand lizard, *Lacerta agilis*, measured as (1) the mean number of disturbances of scale pattern characters and (2) fluctuating asymmetry in scale characters. (Simplified from Zakharov, 1989).

discussed above. There is evidence that FA and morphological variability both increase as organisms are exposed to increasingly stressful conditions. An example from Zakharov's (1989) work with the lizard, *L. agilis*, is provided in Figure 2.4; when FA in the number of lizard scales is plotted against temperature, FA increases away from optimal conditions in a parallel manner to the production of morphological deviants. These and substantial additional data, especially in *D. melanogaster*, indicate that FA is elevated under extreme environmental conditions, particularly when these conditions include temperature extremes.

FA has been shown to vary with environmental conditions in nature. Stressful conditions are often expected at the borders of species distributions where conditions can be often marginal for survival, and several studies have shown that high levels of FA can occur in populations from borders. Again focussing on the lizard, *L. agilis*, Zakharov (1989) compared FA in populations over the major part of its range. Lizards from two high altitude populations at the edge of its range were examined, as well as individuals from several low altitude populations. The two high altitude populations were further apart from each other than from the low altitude populations. Despite this, FA

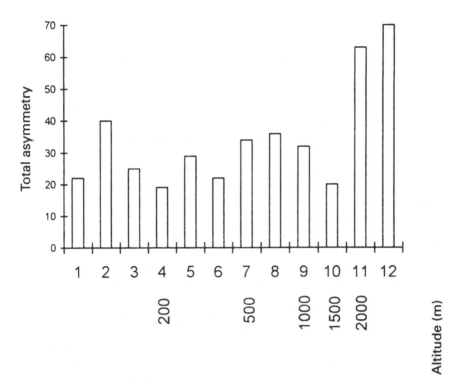

Figure 2.5. Total value of fluctuating asymmetry of eight scale pattern characters in populations of the sand lizard, *Lacerta agilis*. The numbers 1–10 represent different populations of Kuzakhstan; 11 and 12 are populations from two high-mountain regions geographically isolated from each other at the IssykK-Kul' Lake region (11) and Sevan Lake region (12). (From Zakharov, 1989).

based upon eight scale characters was about twice as great in both high altitude populations than in the other populations (Figure 2.5). This suggests that developmental stability was disturbed at the ecological periphery of the range of *L. agilis*. We return to the question of variability in populations from species borders in Chapter 4.

Studies on the phenotypic effects of environmental changes in our own species are restricted in scope compared with those on populations of other organisms. In humans, it is not possible to study mutation and recombination under the extreme stresses discussed above, or to set up comparisons of populations in a range of controlled environments. However, FA has been used quite widely to look for effects of environmental stresses.

Some research suggests FA changes with stress in humans as in other species. An example is a study by Livshits & Kobylianski (1991) on FA in preterm babies, defined as babies born less than 37 weeks after conception. FA was measured for eight morphological traits including the breadth of a baby's hand and its feet. The overall measure of FA based on these traits was negatively correlated with gestational age of the baby, as well as with the health status of both the babies and their mothers. In particular, babies that were extremely premature (26–29 weeks after conception) had very high levels of FA. Cardiovascular and respiratory diseases experienced by babies also had an effect on FA. These associations indicate that FA can provide a measure of the degree of disturbance that infants have experienced during their development.

Another study of FA in humans is described by Kieser (1992) who compared FA in the teeth of children of alcoholic and non-alcoholic mothers. Although children from the alcoholic group did not suffer from foetal alcohol syndrome, they did show increased levels of asymmetry compared to the control group from non-alcoholic mothers. This increased level of asymmetry is likely to be due to stresses experienced by children at the foetal stage. Alcohol consumption in mothers is also associated with other developmental defects such as spontaneous abortion and intellectual disability.

We should emphasize that effects of environmental stresses in humans appear small compared with those obtained in many animal studies. This undoubtedly reflects the fact that levels of stress in most human populations that have been studied are not normally as extreme as those experienced in animal populations. Stress levels in animals can be close to limits for survival.

Levels of fluctuating asymmetry can depend on genotype as well as on the environment. When inbreeding occurs, FA often increases, presumably because inbred individuals are developmentally less stable (Markow, 1995). Moreover, specific genotypes and karyotypes can differ in levels of asymmetry (Parsons, 1990). This is particularly well demonstrated for insecticide resistance in the blowfly, *Lucilia cuprina*. McKenzie & Yen (1995) showed that individuals carrying the gene coding for resistance to the pesticide Dieldrin are more asymmetrical than those without this gene; this effect is apparent even when individuals are cultured under non-optimal temperatures which increase levels of FA in individuals without resistance alleles (Figure 2.6), indicating that genetic effects can sometimes override environmental effects.

The environment and quantitative genetic variation

The above discussion indicates that the incidence of phenotypic abnormalities and fluctuating asymmetry generally increase when organisms experience extreme conditions. We now turn to traits that vary in a continuous or quantitative manner. These include morphological traits such as height, physiological traits such as metabolic rate, and life history traits such as lifespan and development rate.

To look at the effects of environmental conditions on genetic variability in

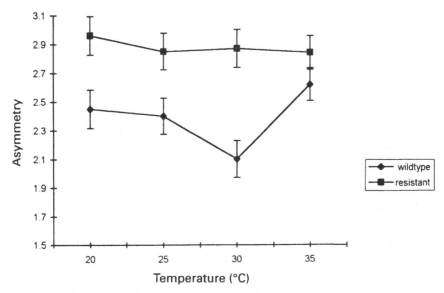

Figure 2.6. Changes in fluctuating asymmetry with environmental temperature in *Lucilia cuprina* blowflies resistant or susceptible to the pesticide Dieldrin. (Simplified from McKenzie & Yen, 1995).

continuous traits, the tools of quantitative genetics can be used. A detailed outline of quantitative genetics is found in texts such as Falconer & Mackay (1995). We briefly need to consider some concepts in this area for the purposes of the following discussion.

The variation in a quantitative trait is measured by its variance, which gives a measure of the spread of a trait around its mean value. If measurements are made on individuals in a population, we are measuring the overall phenotypic variance, designated V_P, which is an assessment of the differences between individuals in a population. For instance, in *Drosophila*, analyses have been carried out on traits such as bristle number, the speed at which males copulate, and the activity of specific enzymes. The phenotypic variance is the total variation in environmental conditions experienced by individuals plus variation between the genotypes of individuals, thus

$$V_P = V_G + V_E,$$

where V_G is the genotypic variance and V_E is the environmental variance. In this equation, it is assumed that the environment and genotype influence a phenotype additively. The genotypic variance can be further subdivided into three components:

(1) The additive genetic variance (V_A) is the component that determines the resemblance between relatives. It is the proportion of V_G that contributes to

similarity between parents and their offspring because the additive effects of genes are passed on from generation to generation.

(2) The dominance genetic variance (V_D) arises when some alleles controlling a quantitative trait show dominance over other alleles. This component contributes to similarity between siblings, but does not influence the similarity between parents and their offspring.

(3) Genetic variance may arise because of interactions between different genes affecting a quantitative trait. For instance, consider a quantitative trait controlled by two loci (A, B) each having two alleles (A_1, A_2 and B_1, B_2). Assume there is no interaction or dominance and alleles have additive effects, and assume that the A_1 and B_1 alleles contribute $+1$ to a trait and the A_2 and B_2 alleles contribute $+2$. The phenotypic values in Table 2.2 will be obtained for the 16 genotypes formed by the 4 gametes. If the alleles occur equally frequently, then the 4 gametes may occur equally frequently if alleles are randomly associated, and the variance will be 1.0. Now assume that the A_1 allele interacts with the B_1 allele so that A_1 contributes 0 to a trait when it is in combination with the B_1 allele. In this case, different values will be obtained as indicated in Table 2.2. The variance will be 2.375 if all the gametes occur equally frequently, indicating the presence of additional variance due to genic interactions (V_I).

We can therefore partition V_G into V_A, V_D and V_I, so that

$$V_P = V_A + V_D + V_I + V_E.$$

In addition to these components, variation between individuals may be affected by 'maternal' effects. These arise when offspring resemble mothers more closely than their fathers. Maternal effects may arise from several causes. One possibility is that the environment experienced by the mother influences the development of her progeny. For instance, in many insects the light-day cycle experienced by the mother can influence the induction of diapause in her progeny. Another possibility is that maternal effects are due to cytoplasmic factors that are passed on through eggs and not via sperm. Mitochondria and microorganisms living in a cell's cytoplasm can be passed on in this manner, and these can influence traits including resistance to environmental stresses.

From the above variances we can obtain three measures of genetic variability.

(1) The degree of genetic determination of a trait defined as its broad-sense heritability (h_b^2),

$$h_b^2 = V_G / (V_G + V_E).$$

(2) The narrow-sense heritability (h_n^2) giving the proportion of the phenotypic variance that is passed on to relatives,

$$h_n^2 = V_A / (V_G + V_E).$$

Table 2.2. *Effect of a genetic interaction on the phenotypes generated by two loci. Numbers represent the phenotypes produced by a combination of the male and female gametes*

In (*a*), the A1 and B1 alleles contribute +1 while the A2 and B2 alleles contribute +2 to the phenotype. In (*b*), the A1 allele contributes 0 when in combination with the B1 allele.

(*a*) No interaction

		Female gametes			
		A_1B_1	A_1B_2	A_2B_1	A_2B_2
Male	A_1B_1	4	5	5	6
gametes	A_1B_2	5	6	6	7
	A_2B_1	5	6	6	7
	A_2B_2	6	7	7	8

(b) Interaction

		Female gametes			
		A_1B_1	A_1B_2	A_2B_1	A_2B_2
Male	A_1B_1	0	3	3	4
gametes	A_1B_2	3	6	6	7
	A_2B_1	3	6	6	7
	A_2B_2	4	7	7	8

(3) The evolvability of a trait (I_A), a measure recently proposed (Houle, 1992) for indicating a trait's ability to respond to selection,

$$I_A = \frac{V_A}{\overline{X}^2},$$

where $\overline{X}$ is the mean of a trait before selection.

There are a variety of ways of measuring heritabilities and variance components in populations. One approach is to carry out directional selection by selecting individuals each generation with high or low values of a trait. The speed of the selection response gives an indication of the narrow-sense heritability of a trait, according to the equation $R=h_n^2 S$. In this equation, R is the response to selection, and S is the 'selection differential' which measures the intensity at which directional selection is carried out. The value of S depends on the amount of variation in a population, expressed as V_P. Because of this, h_n^2 is not a particularly good measure of the extent to which a trait will change when under selection. For instance, a trait may change little under selection when V_P is small even when h_n^2 is close to 1.0. In contrast, I_A does provide an indication of this change because the additive variance is expressed relative to a trait's mean. Larger values of I_A will lead to greater changes in a trait when exposed to directional selection.

Another way of estimating genetic components is to undertake family studies. Estimates of V_A and narrow heritability can be obtained from comparisons between parents and their offspring, while estimates of V_D can be obtained by including comparisons of siblings as well. Finally, components can be estimated by comparisons within and between different strains. If strains are highly inbred (ie., homozygous at most loci), then variation between individuals from the same strain is largely due to environmental effects and can be used to estimate V_E. In contrast, variation between individuals from different strains incorporates both genetic and environmental effects, so that an estimate of V_G is obtainable.

In practice, estimates of genetic parameters are often obtained using several procedures, and combining these can indicate the relative importance of V_D, V_A etc. as well as the reliability of estimates. For instance, Bradley (1978) examined high temperature resistance in the copepod, *Eurytemora affinis*, using comparisons between siblings and between parents and their offspring. The heritability estimates from these procedures differed between the sexes. For females, estimates ranged from 0.11 to 0.40, whereas for males they ranged from 0.72 to 0.89. These estimates indicate that most of the phenotypic variance in males is due to genetic factors, whereas environmental effects are relatively more important for the females. In addition, heritability estimates obtained from the parent–offspring comparisons were similar to those obtained from comparisons of siblings, suggesting that V_D was small because only this component contributes to similarity among siblings. The high heritability estimates for heat resistance suggests that the trait should respond readily to directional selection. This was confirmed by a rapid increase in resistance when a population of *E. affinis* was selected for several generations.

Although these techniques have been widely used to demonstrate genetic variance for responses to stressful environments in the laboratory, extrapolating findings to natural populations can be difficult. One limitation is the nature of the base population upon which experiments are carried out. If results are to be relevant to natural populations, this base population should reflect genetic variation in nature. Unfortunately, genetic components are often estimated with populations that have been kept under laboratory conditions for a long time, or with populations that have been started by only a few individuals from the field. In both cases, only few of the genotypes that occur in nature are likely to be represented in the laboratory populations. A useful way of minimising this problem is to characterize variation within and among 'isofemale strains'. Each of these strains is derived from the progeny of a single female collected from nature. Broad-sense heritabilities can be estimated by comparing differences among and within the isofemale strains.

Nevertheless, even if the different variances can be estimated from populations that accurately reflect genotypes in natural populations, the estimates are still dependent on environmental conditions under which they are obtained. Three examples demonstrate the importance of environmental conditions on levels of genetic variation. The first is a study by Derr (1980) on the American cotton stainer bug, *Dysdercus bimaculatus*, a colonizing insect that feeds on seed crops. A trait directly affecting the fitness of this species is the time taken for a female to lay its first clutch. This trait determines the ability of a female's progeny to use transient seed crops for food because the seeds are only available over a short period. The heritability of the timing of the first clutch was examined under favourable conditions or under stressful rearing conditions when the water available to bugs was restricted. Heritabilities varied from close to 0 under favourable conditions to 0.3–0.4 for the stressful conditions. A large component of the variation in this trait was therefore under genetic control in the stressful conditions but not in the favourable conditions. However, when I_A values are compared, they are around 0.003 in both environments, indicating that the trait will not necessarily change more under stressed conditions.

The second example is from *Drosophila melanogaster* and involves the effects of radiation on genetic variance for longevity, a study carried out over two decades ago (Westerman & Parsons, 1973). In this experiment, four inbred strains were crossed together in all possible combinations including crosses within strains, to give a total of 16 crosses. This type of cross is known as a 'diallel' cross, and it can be used to estimate additive genetic effects, non-additive genetic effects (which include V_D and V_I), and maternal effects. At the extreme dose of 1.2 kGy of radiation, where longevity was around 1/10 of the unirradiated controls, the non-additive component was small but significant, and the additive genetic component predominated (Table 2.3). In contrast, the non-additive variance was more important at lower doses. An estimate of V_E indicated that this component decreased with increasing dose. The example illustrates that the additive genetic variance can change with levels of an environmental stress. In addition, the relative importance of the non-additive and additive components of variance can also change.

Finally, Kawecki (1995) considered heritable variation in life history characters in the cowpea weevil, *Callosobruchus maculatus*, when reared on its normal host (mung beans) versus novel hosts (adzuki beans, red beans and

Table 2.3. *Changes in relative importance of additive and non-additive genetic effects for the longevity of* Drosophila melanogaster *measured at different levels of* ^{60}Co-γ- *irradiation*

Genetic effects were determined from a statistical analysis of the results of crosses between four inbred strains (numbers in the table represent mean squares, obtained from an analysis of variance). The error term represents an indication of environmental effects.

	Radiation dose (Krad)					
	0	40	60	80	100	120
Additive effects	669.7	122.0	54.4	14.1	24.6	143.7
Non-additive effects	380.5	221.9	83.6	52.0	43.7	19.6
Error	150.6	57.4	24.2	12.0	8.1	5.2

Source: Modified from Westerman & Parsons, 1973.

black grams). He found that the narrow- and broad-sense heritability for larval survival was highest on one of the novel hosts (black gram) and low on the original host. In contrast, narrow-sense heritabilities for two other life history traits (development time and growth rate) were relatively higher when weevils were reared on their normal host.

In two of these examples (and in the case of survival in the weevil example), V_A and heritability increased as environmental conditions become more unfavourable. However, it is difficult to make generalizations about changes in heritability and the relative importance of different genetic components (Hoffmann & Parsons, 1991). The additive genetic variance tends to increase with increasing levels of stress in several studies, but there are many exceptions. This may reflect the diversity of organisms, types of stresses and stress intensities that have been used to examine such changes. Nevertheless, it is clear that genetic components of variance often do change with the environment and should not be regarded as constant. At stress levels causing severe reductions in longevity, the additive genetic variance may be particularly important. In *Drosophila*, this has been shown for resistance to desiccation, anoxia, ethanol, ether and other chemicals (Hoffmann & Parsons, 1991).

As well as influencing the different components of genetic variation within populations, environmental factors can influence the nature of genetic differences between populations. This is most evident in the phenomenon of 'heterosis', which often occurs when crosses are carried out between different populations or strains. When a trait shows heterosis, the mean value of the F1 resulting from a cross between two strains exceeds the mean value of both the

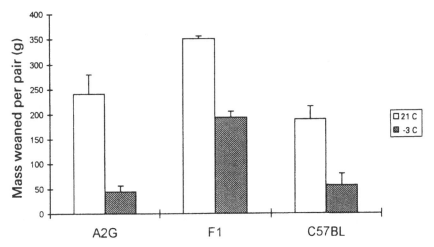

Figure 2.7. Mean mass of young weaned per pair of mice raised by two strains and F1 hybrids at 21°C or −3°C. (After Barnett & Coleman, 1960).

parental strains. Heterosis has been particularly well documented in plants, and is often used to generate plants with superior yield by crossing different varieties of the same species.

In animals, heterosis has also been widely demonstrated, and some experiments show that its magnitude tends to increase as the environment becomes less favourable. We can illustrate this with an example involving mice developing in optimal or cold environments (Barnett & Coleman, 1960). Two inbred strains of mice were crossed to produce hybrids. Crosses within the parental strains were also carried out. Young mice were maintained at an optimal temperature (21°C) or transferred to a cold room (−3°C). Mortality and growth was recorded to the age of 6 weeks. At 21°C, there was some evidence for F1 heterosis in all traits, but the degree of heterosis was much greater at −3°C. The results can be summarized by considering the total mass of the young produced, which incorporates the number of young as well as their size. The data are given in Figure 2.7 and indicate that the F1 scores are about 1.6 times greater than the mean of the parental scores at 21°C. The corresponding figure for the −3°C environment is 3.8, highlighting an increased degree of heterosis in the stressed environment. An increase in heterosis with stress has been demonstrated in several *Drosophila* experiments and in research on agricultural animals. One reason why heterosis is more apparent under stressful conditions is that the effects of deleterious genes are enhanced under these conditions. For instance, Kondrashov and Houle (1994) compared *Drosophila melanogaster* lines which had accumulated deleterious mutations and a line without such mutations. They found that the difference between these lines for

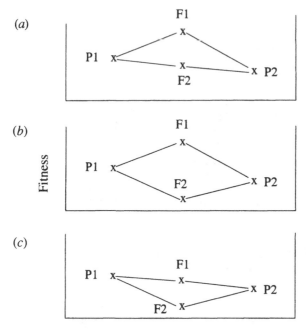

Figure 2.8. Possible outcomes of crosses between two populations (P1, P2), showing (*a*) F1 heterosis without F2 breakdown, (*b*) heterosis and F2 breakdown, and (*c*) breakdown in the absence of heterosis.

fecundity and viability increased dramatically as conditions became stressful.

Finally, the environment may influence the nature of genetic interactions that often arise when crosses are carried out between populations. When two unrelated populations are crossed, the mean value of a fitness-related trait often decreases in the F2 generation below the mean values of the F1 and parental populations. This phenomenon is known as 'F2 breakdown', and is thought to arise from interactions between genes within a population. It is commonly believed that complexes of genes are selected to act together to produce optimal phenotypes in a population. The genes that interact to produce these complexes are expected to differ between populations, because selection pressures will favour genes that interact with other genes in the same population, not in a different population. If genes have been selected to act together within a population, the mean fitness of F2 individuals may be lower than that of both the parental strains, because favourable combinations of genes can be broken up by recombination. This will result in genes from one population being combined with those from the other population, destroying any adaptive complexes that have been built up. The phenomenon of F2 breakdown is illustrated in Figure 2.8 which presents some possible outcomes when populations are crossed.

Some studies on F2 breakdown suggest that it is less likely to be detected under stressful environmental conditions than under favourable conditions. For instance, Tantawy & El-Helw (1970) crossed populations of *Drosophila melanogaster* from Scotland, Japan and Egypt and scored the fitness of individuals at three temperatures (15, 25 and 28°C). The F2s in the crosses were inferior to one or both parents for egg production, emergence and longevity, indicating that F2 breakdown was occurring between these populations. For each trait, F2 breakdown was larger at 25°C, which is close to the optimal temperature for *D. melanogaster*, than at 15°C and 28°C, which are further away from the optimum. However, F2 breakdown may also increase under stress. For instance, in crosses among populations of *Drosophila serrata*, there was increasing evidence for F2 breakdown in development time, viability and fecundity as the environment became less favourable, (Blows & Hoffmann, 1996).

The above findings illustrate that the different types of genetic components affecting quantitative traits can be influenced by the environment. One of these components (heterosis) is more likely to be detected as environmental conditions become less favourable. Generalizations about the other components are less clear-cut, although V_A may be more important under adverse conditions than V_D.

These environmental effects imply that heritabilities will depend on the environment in which they are measured, and that differences between strains under one set of conditions may not be apparent under another set of conditions. This can make it difficult to predict responses to selection in one environment from responses in another environment because the way that genes influence traits depends on the environment in which the traits are measured. If we are to make predictions about evolutionary change on the basis of estimates of genetic parameters, we need to have estimates that are relevant to the environmental conditions being considered.

Ideally, quantitative genetic studies should be carried out under field conditions, whereas until recently most studies have been carried out under laboratory conditions. Field studies of this type are difficult because parents and offspring need to be identified individually in order for comparisons to be made between them. This is only possible for a few species such as some birds where individuals, their mates, and their offspring can be identified in the field. For example, van Noordwijk, van Balen & Scharloo (1988) were able to measure the heritability of body size in great tits, *Parus major*, by measuring the size of individual adults and their progeny. In other organisms, it may be possible to obtain information relevant to field conditions by collecting parents from the field and testing their offspring under laboratory conditions. For example, several researchers have measured body size and other traits in *Drosophila* collected from the field, and have correlated these measurements with those carried out on their offspring reared in the laboratory. This

Table 2.4. *Heritability estimates for heat resistance in* Drosophila simulans *reared under field and laboratory conditions*

	Regression coefficient (standard error)	Probability	Heritability estimate
Estimates for field-reared parents			
mother–son	0.40 (0.10)	0.0008	0.80
mother–daughter	0.35 (0.07)	0.0001	0.70
father–son	0.24 (0.09)	0.0095	0.48
father–daughter	0.23 (0.08)	0.0075	0.46
Estimates for laboratory-reared parents			
mother–son	0.28 (0.11)	0.0092	0.56
mother–daugher	0.21 (0.10)	0.0168	0.42
father–son	0.21 (0.11)	0.0287	0.42
father–daughter	0.17 (0.09)	0.0287	0.34

Source: From Jenkins & Hoffmann, 1994.

approach can be used to obtain field heritability estimates once a number of assumptions are made (Riska, Prout & Turelli, 1989). As an example, Table 2.4 shows heritabilities estimated for the heat resistance of *Drosophila simulans*, based on comparisons of field-caught flies and laboratory-reared individuals. The comparisons of field flies and their progeny indicate that there is heritable variation in the field population because the heritability is significantly greater than zero. In addition, there seems to be a maternal effect induced by field conditions because laboratory progeny are more similar to their field mothers than to their field fathers.

While these types of studies provide more meaningful estimates of genetic parameters than laboratory studies, they are still limited because estimates of variance components under one set of field conditions may differ from those obtained under another set. We can illustrate the difficulties that can arise when attempting to extrapolate from one set of field conditions using the study of great tits by van Noordwijk *et al.* (1988) mentioned earlier. Heritability estimates for body size based on correlations between parents and offspring were fairly large (around 0.5) under favourable conditions. However, this value decreased to zero when nestlings were raised under poor feeding conditions which led to considerable nestling mortality (>25%). This does not necessarily mean that the heritability of body size was lower under stressful conditions. It may simply reflect the fact that genes affecting size under stressful conditions are different to those influencing size under non-stressful conditions.

Ideally, we should always attempt to obtain heritability estimates for traits under conditions corresponding to those when the traits are under intense selection. This will often mean obtaining estimates under stressful conditions in nature .

Selection under extremes: does heritability always matter?

Short-term responses in traits can be predicted by estimates of heritability and other measures of genetic variability. However, it is not clear if such estimates are always meaningful under extreme conditions, particularly in the case of traits that are predominantly canalized and have low plasticity.

As mentioned earlier, the expression of morphological traits showing little variability can be influenced by stressful conditions; extreme conditions often lead to the appearance of aberrant phenotypes. In *Drosophila*, phenotypes that only arise under stress may be similar to those produced by mutant genes. For instance, when eggs of *Drosophila melanogaster* are exposed to ether vapour in sub-lethal doses, a few of the survivors develop an abnormal phenotype known as 'bithorax', where the thoracic region of the insect is duplicated (Waddington, 1957); this phenotype is also the result of a mutant gene. Phenotypes produced by extreme conditions that resemble those produced by mutations are often referred to as 'phenocopies'.

Waddington (1957) demonstrated how the frequency of aberrant phenotypes could be increased by selection within a strain of flies. For instance, in the case of bithorax, he used abnormal flies generated by ether as parents in the subsequent generation, and repeated the ether treatment on their eggs. When this selection procedure was repeated for several generations, the incidence of bithorax increased steadily, and the selected strain eventually produced the bithorax condition without exposure to ether. The bithorax characteristic had therefore been altered from being induced by the environment to becoming a genetically determined condition. Waddington referred to this process as 'genetic assimilation', whereby a variant that is originally only expressed as a consequence of a stress can become assimilated and expressed in the absence of the stress. A trait that was initially canalized was therefore altered by exposure to stressful conditions. This process involved genes present in the base population rather than newly arisen genes generated by mutation, because an increased expression of aberrant phenotypes was not observed when selection was carried out on inbred lines. These lines lack genetic variability so that any genetic changes within lines would mainly be due to mutation.

The genetic assimilation process may apply generally to canalized traits whose expression is relatively constant under normal conditions, ie. show a low level of plasticity. In the absence of a stress, an organism is expected to be buffered against environmental changes, in that the organism produces the same phenotype under a wide range of conditions. However, the buffering action of genes that control development can be opposed by stressful conditions, changing the phenotype and exposing genetic variants that were formerly hidden to selection.

Similar experiments have been undertaken with aberrant phenotypes induced by other severe environmental stresses. For example, a short exposure to high temperature applied at a critical stage in the development of pupae causes the appearance of the 'crossveinless' phenotype, involving the absence or reduction of crossveins that are normally present in the wings of *Drosophila*. Selection for crossveinless flies by Milkman (1960) ultimately produced strains with large numbers of flies expressing this phenotype in the absence of the stress. The process of genetic assimilation can therefore also occur with stresses likely to be encountered in nature. Some attempt has been made to identify the genes that control the appearance of the crossveinless phenotype. By using a complicated series of crosses, it was shown that only a few genes were involved, and that these genes had a large effect on the crossveinless phenotype. These genes were shown to exist in natural populations of *D. melanogaster*, indicating the potential for genetic assimilation in natural populations.

We can summarize the results of these early experiments with *Drosophila* in terms of four important points.
(1)   The stimulus that triggers the abnormality is highly specific; in the examples considered, eggs exposed to ether and pupae exposed to high temperatures at critical stages produced bithorax and crossveinless respectively.
(2)   As a consequence of abnormalities, variability in traits with low plasticity can be substantially increased in sub-lethal environments.
(3)   Severe stresses that trigger abnormalities can reveal genetic variation which is not normally expressed because the abnormalities could otherwise not be selected.
(4)   The production of abnormalities appears to be mainly controlled by major genes.

The mechanism underlying this process has not been clarified. Waddington argued that genes directly affecting the canalization of traits were important. In his view, different genes produced the same phenotype prior to selection because of the presence of genes controlling canalization. The latter were reduced in frequency during selection, exposing genes that had previously produced the same phenotype to selection. However, the genetic assimilation process may also be explained by a simple threshold model (Bateman, 1959). Assume that the tendency to produce an aberrant phenotype is determined by a normal distribution, but the phenotype only appears once a threshold is exceeded. Without stress, this threshold is rarely exceeded as in Figure 2.9(*a*).

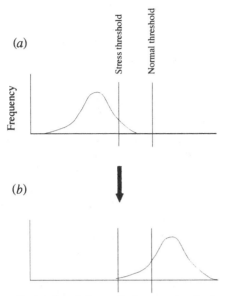

Number of genetic factors affecting expression of variant

Figure 2.9. A simple threshold model to account for genetic assimilation. Initially in (*a*), a phenotypic variant is only expressed under stressful conditions when a threshold is exceeded, but not under normal conditions associated with a different threshold. Eventually in (*b*), the trait is expressed under normal conditions because genes have accumulated during selection for the trait under stressful conditions.

However, assume that a stress shifts the threshold as indicated in Figure 2.9(*b*), resulting in the expression of the aberrant phenotype. Selection causes an increase in the frequency of alleles contributing to this phenotype, eventually shifting the threshold to the right and resulting in the expression of the phenotype even in the absence of stressful conditions. Rendel (1967) proposed a somewhat more complicated threshold model to account for the invariant nature of many traits.

Recent molecular studies by Gibson & Hogness (1996) on the *Ubx* gene associated with the bithorax condition have provided data on the types of allelic changes that might be involved in such models. Selection was undertaken to increase the production of bithorax following an ether stress. This resulted in lines producing bithorax phenocopies at a frequency of 45% after six generations, compared to 13% in the starting population. By extracting DNA from the lines and screening for variation in the *Ubx* gene, Gibson & Hogness (1996) showed that allelic changes had occurred and involved a site within this gene. The ether treatment therefore exposed variation in the *Ubx* gene to selection.

However, a simple threshold model may be too simple for other invariant traits. There is evidence that the environment or major mutants that upset development may act via different developmental systems that are partly independent (Scharloo, 1991). This means that changes in a trait along a single axis may not always be realistic.

In *Drosophila*, canalized traits other than those associated with morphology may show little evolution unless conditions are stressful. An example is described by Neyfakh & Hartl (1993) who considered embryonic development in *Drosophila melanogaster*. They pointed out that the rate of development of embryos is relatively constant at a particular temperature, and that attempts to change this rate by selection have been unsuccessful. However, by selecting for increased development at 32°C, which is close to the lethal limit of *D. melanogaster*, Neyfakh & Hartl obtained an increase in embryonic development rate in two lines with a realized heritability of 5–9%. In one of the lines, development rate also showed a concomitant increase at 25°C, which is close to the optimal temperature for this species and suggests that development was generally altered.

There is evidence that stressful conditions can result in specific types of responses to selection for life history traits. In *D. melanogaster*, Robertson (1964) found that selection for larger size on culture medium deficient in protein did not change larval development time. However, when medium was deficient in RNA, there was a change in the larval period, indicating that different genes were involved under these conditions. Changes in size may, like morphological abnormalities, be subject to the process of genetic assimilation; Robertson (1964) found that while responses to selection for size tended to be diet-specific early in the selection process, this was no longer the case in later generations.

The effects of stress on evolutionary change have also been investigated in other organisms. In plants, aspects of floral morphology can be canalized, and genetic variability in floral traits may be released under stress. For instance, in the genus *Linanthus*, flowers normally have five corolla lobes, but there is a low incidence of flowers with fewer or more lobes in natural populations (Huether, 1969). This incidence increased under stressful conditions, resulting in small plants that bloom late and produce few seeds because of changing daylengths and higher temperatures. Huether (1968) showed that there was genetic variation underlying differences in corolla lobe number, by selecting lines under unfavourable conditions for increased and decreased lobe number. Differences between the selected and control lines were larger under stressful conditions than under optimal conditions.

There is evidence that the presence of major mutations can have similar effects to stresses in influencing responses to selection. In mice, the numbers of whiskers (known as vibrissae) have been selected by Kindred (1967). The number of vibrissae is usually invariant in populations, varying rarely from 19.

However, in mice carrying a mutant gene known as Tabby, there is more variation in vibrissa number, presumably because the presence of this gene or a linked gene influences some component of development associated with the production of whiskers. When normal mice were subjected to selection for increased or decreased vibrissa number, there was no change in this trait after 10 generations. However, in the presence of the Tabby allele, selection was successful, particularly for low vibrissa number, and the resulting heritability was in the order of 30–50%. Hence the presence of a major gene upsetting development increased variability and resulted in a selection response. There are other examples of mutants increasing phenotypic variability and fluctuating asymmetry in traits, where the variability can subsequently be selected (Scharloo, 1991).

Can selection for these major pleiotropic changes and the genetic assimilation process be regarded as a model for adaptive evolution for invariant traits? At present we do not have an answer to this question. As already mentioned, genes influencing crossveinless in *Drosophila* are found in nature. These genes are quite common, suggesting that they could be favoured in some habitats. However, phenotypes such as bithorax that are generated by stress exposure are unlikely to survive in nature. It is possible that selection for abnormalities can lead to major evolutionary changes, but appropriate conditions may be rare, making them difficult to study. Nevertheless, if variation in canalized traits that are buffered during development does not appear unless conditions are stressful, it is difficult to see how much evolutionary change will occur in the absence of stress. The role of stress-induced variation in the evolution of traits other than those involving morphology also deserves attention. We return to the issues raised here in Chapter 6.

Summary

Recombination frequencies and mutation rates tend to increase under adverse conditions, particularly when conditions are close to lethality. Natural selection may adjust both mutation and recombination rates to intermediate levels, but it is not clear if the increase in the rates of these processes under stress is adaptive. While stressful conditions can lead to a marked increase in the mutation rate in bacteria, more investigations are needed on the issue as to whether this increase is random or specific to loci that are under selection.

Phenotypic variability of many traits increases under stress. Stress applied at critical developmental stages can lead to abnormal adult phenotypes. Developmental instability, particularly when expressed in terms of fluctuating asymmetry, increases under stressful conditions in both the laboratory and in natural populations.

The extent to which variation in a trait is genetically determined can change with environmental conditions. Heterosis in crosses between populations and F2 breakdown can also be influenced by the environment. These conclusions are important for the application of quantitative genetic models to natural populations.

Many traits appear invariant unless conditions are stressful, or unless there are genes with major effects on development segregating in populations. The process of genetic assimilation results in environmentally-induced variation becoming expressed under non-stressful conditions and may serve as a model for the evolution of invariant traits, although it is difficult to determine how important this process has been in natural populations.

CHAPTER 3

Natural selection in extreme environments

Here we consider evidence for natural selection in extreme environments. We start with a brief overview of traditional methods for demonstrating selection, before considering selection in response to stresses arising from human activities. These have produced many well-known case studies of natural selection. We then examine selection responses to climatic extremes unrelated to human influences.

This chapter also addresses the effects of extreme environments on fitness differences among genotypes. We consider the case of genetic differences affecting enzymes; these tend to affect fitness when organisms experience stressful conditions. In addition, associations between fitness and the overall level of heterozygosity of an organism at enzyme loci become more evident when environments are unfavourable. We also examine the way selection acts under extreme conditions. Recent techniques for describing the effects of selection and identifying targets of selection are briefly outlined. These techniques are starting to be applied to responses to environmental change. Early findings suggest that the types of selection operating on traits can change between years, and that selective factors acting within populations may not be the same as those acting between populations.

Finally, we address the importance of extreme environments in adaptive responses. There has been an ongoing debate in evolutionary biology for many years about whether adaptation involves a few genes with large effects or numerous genes with small effects on traits. Although much of the literature is concerned with gene number under optimal rather than extreme conditions, there is evidence that major genes have a role in adaptive responses when conditions change drastically. A simple genetic basis involving one or a few major genes might therefore indicate that intense selection has occurred under extremes, whereas the involvement of many genes is more consistent with gradual environmental changes. These and other predictions remain to be evaluated in natural populations.

Demonstrating natural selection: traditional approaches

Endler (1986) reviewed traditional methods for demonstrating selection and these are summarized in Table 3.1. The optimality methods and comparative method (9, 10) are considered in a later chapter (Chapter 5). These are both indirect methods for demonstrating selection based on differences between species, whereas this chapter focusses on phenotypic variation within species and the genetic basis of this variation. Phenotypic variation within a species may involve traits that vary in a continuous manner, such as size measurements, enzyme activity or stress resistance. It may also involve traits that mainly vary in a discrete or discontinuous manner, such as body markings or body colour. The different forms of traits that vary discontinuously are usually referred to as 'morphs'. Populations that contain more than one morph are therefore referred to as 'polymorphic', as opposed to 'monomorphic' populations without discontinuous variation in a particular trait.

The first method in Table 3.1 involves associating variation of a trait with variation in the environment over a large area, often encompassing the entire distribution of a species. Where species occur in a variety of climatic regions, associations can be based on a wide range of conditions. This method is one of the most commonly used for demonstrating natural selection. Its main limitation is that associations may not reflect a causal relationship between variation of a trait and the environment. If different values of a trait are associated with dry and humid areas, it does not necessarily follow that humidity selects directly on this trait. A factor related to humidity might be the selective agent. For example, humidity often tends to be negatively correlated with high temperature because warm areas tend to be dry, and trait values favoured by high temperature may be common in dry areas. In addition, associations may not reflect causal relationships because the effects of an environmental variable are indirect. Climate can influence the density of predators or parasites and these biotic factors may in turn impose selection on the organism under study. Additional information is almost always required before a causal role can be attributed to environmental variables from these types of associations.

A related approach (method 2) is to correlate variation in a trait with environmental changes over a short distance. This approach has been applied to stresses that appear abruptly such as sudden changes in the concentration of heavy metals in soils. Associations have also been based on environmental variables that vary over time rather than in space (method 3). For example, seasonal climatic changes can be associated with the frequency of morphs in populations. Since associations between traits and environmental variables can arise by chance rather than from selection, findings obtained by any of

Table 3.1. *Methods for detecting selection in natural populations.* (Modified and abbreviated from Endler, 1986)

Method	Hypothesis assuming no selection	Selection hypothesis
1. Association with environmental factors at the geographic level	No association between traits and environmental factors	Geographically varying selection results in association
2. Association with environmental factors at local level	No association between traits and local environmental variation	Local selection results in association with traits
3. Association with seasonal environmental changes	No seasonal association	Seasonally-varying selection on traits
4. Long-term studies of trait distributions	Random changes in traits over time	Stability or regular directional change in trait due to selection
5. Perturbations of natural populations	Random changes after perturbation	Trait distributions change after perturbation due to selection
6. Predictions based on fitness tests	Predictions on the basis of fitness tests not met	Predictions based on fitness tests met
7. Comparisons among cohorts, age classes or life cycle stages	No consistent change in traits between age classes	Selection causes changes between age classes
8. Predictions based on prior knowledge of physiology, biochemistry etc.	Predictions based on prior information met	Predictions based on prior information not met
9. Predictions based on optimization	Optimization predictions not met	Optimization predictions met
10. Comparisons between species	No association between traits and environments	Association between traits and environments
11. Deviations from null models	No selection in models	Selection causes deviation from models

these methods should ideally be replicated. Environmental associations over a large area can be replicated by testing for the same associations on different continents, while seasonal associations can be replicated by collecting data over several years.

In the fourth method, natural selection is implicated by long-term stability of traits or by long term directional changes in traits. If there is variation in a quantitative trait but its mean value does not change over a long period of time, then natural selection may be continuously favouring the same optimum value of the trait. Alternatively, a continuous change in the mean value of a trait in the same direction can indicate that selection is favouring values on one side of a trait's distribution. These approaches are useful if environments are stable or continuously changing in the same direction.

The fifth method involves following the distribution of a trait or the frequency of morphs in a population. Morph frequencies can often be changed artificially by removing or introducing individuals. It is also possible to test if morph frequencies or mean values of a trait change when a population is exposed to environmental changes. These may be artificially induced or may occur naturally as a consequence of unusual conditions. Measurements of traits or morph frequencies before and after the perturbation can provide evidence for selection. Periods of environmental stress provide particularly good opportunities to study natural selection using this method.

Information on the fitness of individuals is used in the sixth method. For example, large individuals may have a higher reproductive success than small individuals, suggesting that natural selection will favour large size. In a related method, comparisons are carried out between different age classes or different stages in an organism's life cycle (method 7). For example, in insects, trait distributions can be compared between eggs, larvae, pupae and adults. Selection is suggested when the distribution of a trait or the frequency of a morph changes between classes or stages.

Endler (1986) distinguishes two methods that make predictions about the way traits or morphs are distributed in populations as a consequence of natural selection. The first (method 8) predicts the types of changes that might occur in a population on the basis of prior knowledge about how selection is likely to influence a trait or morph. This may take the form of prior fitness estimates (method 6), or prior information about traits, especially those involving energetics, physiology or behaviour. For example, if a morph decreases the energy demand of an organism then it is expected to increase when there is a food shortage. The second of these approaches (method 9) is the optimality approach discussed in Chapter 5. Predictions about the distribution of a trait in a population are made by assuming that the trait has been under selection for a long time.

The final method depends on the identification of genotypes underlying traits under selection. The way these genotypes are distributed in a population can be compared to predictions from models that assume no selection, known as 'null models'. This approach is not relevant to our discussion because selective agents are not considered.

It should be emphasized that changes due to selection will only occur when variation in traits or differences between morphs have a heritable component. The effects of selection would otherwise not be passed to the next generation, precluding evolution. It is not sufficient to demonstrate that individuals with one morph or trait value had a greater reproductive success than those with another one, because the frequency of the morphs or the distributions of traits will not change unless they are transmitted to the next generation.

Extreme environments generated by human activities

Human activities are becoming more widespread every century. There are now many well-studied examples of these activities causing natural selection. Selection can often be intense as organisms respond to novel conditions.

One of the best-known examples is adaptation by plants to waste from mining operations. Waste is often discarded as tailings containing high concentrations of heavy metals. Most plants cannot grow on tailings, but a few species have colonized such areas by evolving resistance to heavy metals. To test whether plants are resistant, seedlings are grown in solutions with heavy metals, and the degree of root development is measured as an assessment procedure. Such experiments have demonstrated the existence of metal-resistant and sensitive populations of the same species. For instance, Figure 3.1 shows the growth of three populations of a grass, *Deschampsia cespitosa*, on solutions containing copper and nickel (Cox & Hutchinson, 1980). One population came from a nickel/copper contaminated area, within 1.6 km of a smelter in Ontario, Canada. The other populations were from two uncontaminated sites, 150 km from the smelter and in an English pasture. The data clearly indicate that plants from contaminated soil are relatively more resistant.

There are many other examples of resistance to heavy metals evolving, including cases where resistance levels have changed over short distances. For instance, the classic work of Hickey & McNeilly (1975) on mine tailings in Wales indicated changes in resistance over a few metres. The sharp changes in resistance reflect the strength of selection caused by heavy metals. Plants that do not have a resistant genotype simply cannot grow on contaminated soils. Conversely, the low level of resistance in uncontaminated environments suggests that there are costs associated with resistance. There is direct evidence that plant genotypes adapted to contaminated soils are not successful elsewhere. We return to the issue of costs in the next chapter.

Heavy metal studies involve method 2 because genetic differences have been found over relatively short distances. They also involve method 6 because the metal resistance is measured using a trait (growth rate) closely related to fitness. Although the selective agent appears to be obvious in such studies, it can still be difficult to identify causal associations between traits and environ-

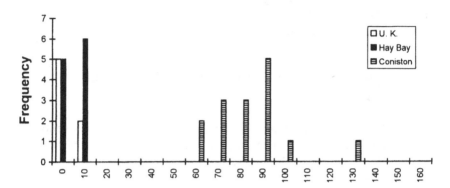

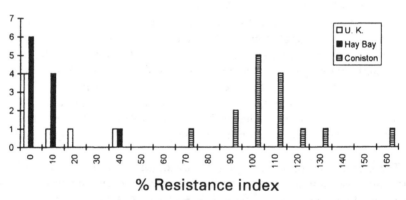

Figure 3.1. Distribution of copper and nickel resistance in clones from three populations of *Deschampsia cespitosa*. Clones were obtained from tillers and seed collected near a copper/nickel smelter at Coniston, Ontario. Clones were also obtained from an uncontaminated site at Hay Bay, Ontario, and an uncontaminated site in Derbyshire, U.K. (Simplified from Cox & Hutchinson, 1980).

mental factors. Mine tailings and uncontaminated soils can differ for other environmental variables such as nutrient levels that tend to be low on tailings.

Adaptive changes have occurred in response to other chemical stresses associated with human activities. Numerous insect species have developed strong resistance to insecticides applied over large areas to control these pests. Similarly, many weeds have evolved resistance to herbicides. In both cases, chemicals impose severe stresses on target organisms because farmers aim for their complete control.

Not all adaptive responses to human activities involve chemicals. For instance, water discharges from a nuclear power plant have led to genetic changes in heat resistance in the copepod, *Eurytemora affinis* (La Belle & Bradley, 1982). Progeny of samples collected near the power plant, where copepods are exposed to rapid temperature increases, recover more rapidly from a heat shock than progeny collected further away from the plant in areas not affected by discharges. This indicates selection for increased heat resistance close to the power plant.

The copepod study is unusual in that it compares nearby populations, whereas most animal comparisons involve populations that are widely separated geographically. Even in experiments with heavy metal resistance in small invertebrates living on metal-contaminated sites, comparisons are usually made between strains from contaminated sites and those from a distant reference site that is uncontaminated (Posthuma & Van Straalen, 1993). The high levels of movement characteristic of many animals has the potential to prevent genetic differences building up over short distances. However, the copepod study and a handful of other studies show that genetic differences can still develop over short distances if selection is intense.

Melanism in the peppered moth, *Biston betularia*, remains one of the classic cases where selection on a genetic polymorphism has been associated with human activities (Brakefield, 1987). Although this moth is normally light in colour, a dark melanic form (*carbonaria*) is common in some areas of England. The frequency of melanics increased in the 1800s at the expense of the non-melanic *typica* form as the effects of industrial pollution became widespread.

This genetic change did not occur as a direct consequence of increased pollution levels, but seems to be largely related to altered levels of bird predation. The *typica* form is normally camouflaged when it settles on tree trunks covered with foliose lichens, and is particularly well-camouflaged when moths are paired for mating. However, industrial pollution caused a decline in lichen numbers and led to the deposition of soot on tree trunks. Many experiments with dead moths pinned on these trunks have indicated that the *typica* morph is much more susceptible to predation by birds than the *carbonaria* morph.

The importance of industrial pollution as a selective agent on this polymorphism is also highlighted by recent changes in the frequency of the melanic morph. A reduction in smoke and sulphur dioxide emissions since the 1960s is associated with a continuing decrease in the frequency of the *carbonaria* morph (Figure 3.2). However, the decrease in *carbonaria* occurs some time after a reduction in pollution levels. This suggests that pollutants are having an indirect effect on the frequency of the morphs. One explanation for this lag period is that foliose lichens providing camouflage for the *typica* morph take time to recolonize areas used by moths for resting.

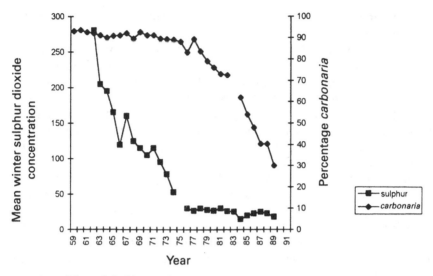

Figure 3.2. Changes in the frequency of the *carbonaria* form of *Biston* and in the concentration of sulphur dioxide in West Kirby, England, from 1959 to 1989. (After Clarke, Clarke & Dawkins, 1990).

Factors other than predation may also influence this polymorphism. Mani (1990) has shown that some form of non-visual selection is needed to account for the geographic distribution of the morphs as well as changes in morph frequencies. This other selective agent has not been identified. However, in species such as ladybirds, melanics may be selected directly by changing pollution levels. Melanic ladybirds may have an advantage over non-melanics when sunshine levels are low because they heat up more quickly and can therefore become active earlier in the day (Brakefield & Lees, 1987). This would enable them to spend more time searching for food and mates.

Melanism in *Biston betularia* illustrates how the genetic constitution of a population can be altered when the effects of an environmental change are mediated via a biotic interaction. Several methods listed in Table 3.1 have been used to infer selection. Associations were established with environmental variation at a geographic level (method 1) and via long-term environmental changes (method 4). Fitness tests were used to establish the nature of selection acting on the morphs (method 6).

In summary, many cases of natural selection in plants and animals are associated with the direct and indirect effects of humans. The stresses involved tend to be extreme, and the fact that organisms have successfully countered them indicates that rapid genetic changes can occur in these situations. However, we should emphasize that many species have failed to adapt to our activities, as evident by the numerous populations and species that have become extinct following habitat destruction and the introduction of exotic plants and animals.

Selection under climatic extremes

Several genetic polymorphisms that have been under scrutiny for many years have been associated with environmental extremes. Variation in the banding patterns and colour of snail shells have been widely investigated, particularly in terrestrial snails of the genus *Cepaea*. The different banding and shell morphs have long been favourite subjects of research in ecological genetics, because the morphs are genetically determined and morph frequencies can be easily scored in the field. While there is evidence that predation can influence the frequency of banding and colour morphs, associations with climatic variables are also important.

The frequencies of shell colour and banding morphs in both *Cepaea nemoralis* and *C. hortensis* are associated with climatic variables (temperature, shading, humidity), particularly over large areas that include a range of climatic conditions (Jones, Leith & Rawlings, 1977). Shell morphs differ in their absorption of solar radiation. Shells may be dark because they are heavily banded or because of their colour. Field measurements of shell temperatures indicate that dark shells are almost always hotter than light shells. Dark morphs may therefore be relatively less resistant to high temperatures.

In agreement with this hypothesis, temperature extremes have been related to morph frequencies, particularly in habitats at the margins of the distribution of these species. In particular, the brown morph of *C. nemoralis* is abundant in marginal sites with low night temperatures, while mid-banded forms are common in open habitats exposed to temperature extremes (Bantock & Price, 1975). Direct evidence that banding and colour morphs can differ in fitness at high temperatures was obtained by Richardson (1974), who examined morph frequencies in a *Cepaea* population in sand dunes at the time of a high temperature stress. He found that numbers of unbanded and yellow forms among the dead snails were lower than expected, suggesting that frequencies had been altered by natural selection because unbanded and yellow morphs were more resistant to high temperatures than pink and banded morphs (Table 3.2).

The *Cepaea* work has used a number of methods from Table 3.1 to implicate climatic selection. Broad geographical surveys provided some evidence (method 1). Physiological information was used to make predictions about likely associations between morph frequencies and environmental factors (method 8). Finally, a natural perturbation provided direct evidence for natural selection via heat stress (method 5).

Morphological variation in other species of snails has also been related to

Table 3.2. *Distribution of shell morphs in dead* Cepaea nemoralis *collected from sand dunes at two sites*

Numbers in brackets indicate expecteds based on a random sampling of live snails

	Shell colour			
Banding pattern	yellow	pink	brown	Total
Site 1				
unbanded	0 (0.5)	3 (10.5)	12 (1.3)	15 (22.3)
1 band	12 (12.7)	78 (81.1)		90 (93.8)
5 bands	10 (17.6)	130 (111.2)		140 (128.8)
Total	22 (30.8)	211 (202.8)	12 (11.3)	245
Site 2				
unbanded	5 (15.2)	27 (45.6)	1 (0.0)	33 (60.8)
1 band	10 (21.3)	291 (258.1)		301 (279.4)
5 bands	4 (21.3)	169 (145.5)		173 (166.8)
Total	19 (57.8)	487 (449.2)	1 (0.0)	507

Source: From Hoffmann & Parsons, 1991, after Richardson, 1974.

environmental conditions. For example, brown morphs of the intertidal snail, *Nucella lapillus*, tend to occur on exposed shores, whereas white morphs predominate on protected shores where high temperatures and desiccation stresses are more common (Etter, 1988). The brown morphs suffer higher mortality in the protected environment than white morphs because they heat up faster and dry out more rapidly.

Selection on other polymorphisms has also been associated with climatically extreme conditions. As mentioned in the previous section, resistance to pesticides provides some of the best examples of rapid evolutionary changes in natural animal populations. While alleles coding for resistance have extremely high fitness when a pesticide is being applied, the same allele may be at a disadvantage under other conditions. Increasing evidence indicates that such a disadvantage is more likely to occur under climatically stressful conditions than under favourable conditions. We can illustrate this in the case of resistance to the pesticide Dieldrin in the sheep blowfly, *Lucilia cuprina*. When this species overwinters at the larval stage, mortalities can be extremely high, with less than 10% of larvae surviving compared to 50–80% at other times. McKenzie (1990) crossed a resistant strain with a susceptible one and scored the number of F2 progeny that carried the resistance gene. In an overwintering population, this proportion was low (9–15%), compared to a much higher frequency (43–53%) at other times and in laboratory tests. The effect of the resistance gene on fitness was therefore enhanced dramatically when tests were undertaken at climatically stressful times. Similar results have been obtained for genes coding for resistance to other insecticides.

The abnormal abdomen polymorphism in the fly, *Drosophila mercatorum*, represents an example of selection for evasion of extreme climatic conditions via life history traits (Templeton *et al.*, 1990). Individuals with abnormal abdomens (referred to as **aa** individuals) carry insertions in the DNA region coding for ribosomal genes. This region is located on the X chromosome. A separate X-linked locus also controls the expression of the **aa** phenotype. Females that are **aa** have abnormal abdomens only when they develop under some environmental conditions, but more importantly they always have a prolonged larval development stage compared to other individuals (non-**aa**) which lack the insertions. In addition, **aa** adults reach reproductive maturity earlier than non-**aa** flies. This in turn results in an increased rate of egg production when females are young, although **aa** females live for a shorter time than non-**aa** females. Templeton *et al.* (1990) showed that these differences were evident under field as well as laboratory conditions. They collected flies emerging from cactus rots in the field, and found that females carrying the abnormal abdomen allele emerged later than wild type females. In contrast, there was no difference in the emergence time of the males, consistent with the fact that the **aa** phenotype is supressed by the Y chromosome in males. This suggests that **aa** females emerge later under field conditions.

These phenotypic differences between **aa** and non-**aa** flies have been related to natural selection by climatic stresses in Hawaiian populations of *D. mercatorum*. Adults normally die within a few days under dry conditions in the field because of desiccation stress, but live much longer under humid conditions. As a consequence, dry conditions are expected to select for a different phenotype than humid conditions. Individuals that mature early as adults and have a high initial output of eggs evade a desiccation stress and will therefore be favoured under dry conditions. This means that **aa** flies should be selected over non-**aa** flies. In contrast, non-**aa** flies will be favoured under more humid conditions because of their increased lifespan and decreased development time, meaning that they will emerge earlier as adults and produce eggs over a longer period of time.

This helps to explain changes in the frequency of **aa** X chromosomes in *D. mercatorum* in response to climatic variation. Flies collected by Templeton & Johnston (1988) along an altitude gradient were scored over a period when there were climatic changes (Figure 3.3). In the study site, dry conditions normally occur at low altitudes and humid conditions are the norm at high altitudes. In 1980, **aa** was relatively more common in the low altitude site (IV) as expected. During a dry year (1981), the frequency of **aa** increased at all sites except IV, while the reverse occurred in a wet year (1982). The weather was

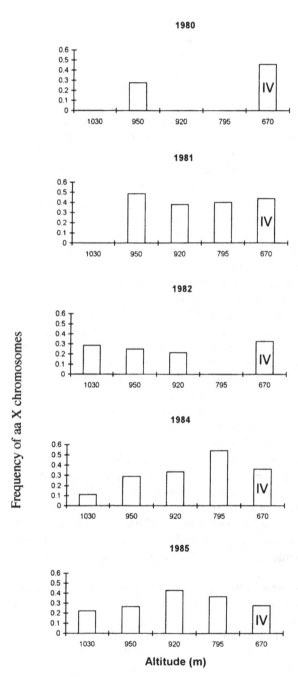

Figure 3.3. Frequencies of X chromosomes with expression of abnormal abdomen (**aa**) in *Drosophila mercatorum* collected over 5 years from locations at different altitudes. (Plotted from data in Templeton & Johnston, 1988).

normal again in 1984 and 1985 except for high humidity at site IV, and this was reflected in changes in the frequency of **aa**. The phenotypic effects of abnormal abdomen on life history traits could therefore be related to selection for evasion of climatic stresses.

The *D. mercatorum* study used several of the methods listed in Table 3.1 for detecting natural selection. The correlation between **aa** frequency and climatic conditions at different altitudes uses the first method. The predictions of changes in **aa** frequency on the basis of the effect that this polymorphism has on lifespan and maturation rate involves the eighth method. Finally, the changes in **aa** frequency at a site following a stress provide evidence for natural selection based on the fifth method because the stress periods represent natural perturbations. The *D. mercatorum* study represents one of the few cases where the effects of selection are understood at the molecular level, and where a causal relationship between genetic changes and environmental changes has been established. It shows the advantages of carrying out simultaneous field and laboratory studies, for which the genus *Drosophila* is particularly useful.

The above consider specific genetic polymorphisms. Is there evidence for selection on quantitive traits being imposed by climatic extremes? Populations are expected to diverge genetically if they experience different environmental conditions. If this occurs in response to extremes, it should be detectable when comparing populations.

There is ample evidence that geographic variation exists for responses to extremes, but such variation is often difficult to relate to natural selection under extremes. We can use an example to illustrate the types of problems that arise in these studies. In mosquitoes, adults have stringent humidity requirements and can only survive when humidity is relatively high. This implies that susceptibility to water loss plays a major role in limiting the distribution of mosquito species. *Aedes atropalpus* is a mosquito with a wide distribution in North and Central America. This species breeds in rock pools, resulting in populations that are distributed irregularly depending on where rock pools occur. Machado-Allison & Craig (1972) tested the mortality of females from four strains of *A. atropalpus* under different levels of desiccation stress (Figure 3.4). The Utah strain was the most resistant to desiccation and came from an arid area in eastern Utah. The Texas strain originated in a semi-arid area and was the next most resistant. The Bass Rock strain was from Massachusetts where the climate is humid, while the San Salvador strain was from a tropical climate which is also humid. The latter two strains were the most sensitive to desiccation even though they originated from very different latitudes. Strain differences in desiccation resistance were therefore consistent with expectations based on climatic conditions.

This study represents an example of method 1 (Table 3.1) where evidence of natural selection on desiccation resistance is obtained from a broad geographic comparison. The results suggest selection by climatic conditions, but the experiments have two limitations in testing predictions about extreme

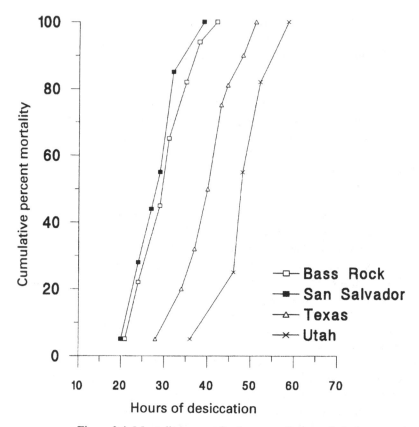

Figure 3.4. Mortality curves for four populations of *Aedes atropalpus* mosquitoes after exposure to <10% relative humidity. (From Machado-Allison & Craig, 1972).

environments. The first of these concerns the desiccation stress used in the experiments (<10% relative humidity for several hours) which may never be encountered in the field. This raises the question of whether mortality selection would ever occur. It seems more likely that short periods of desiccation will select for other resistance traits, such as the ability of the mosquitoes to recover their fertility or fecundity after they have been stressed. The second limitation is that experiments were undertaken with one strain from each location. Although differences among the strains were probably heritable because strains were bred under the same conditions, experiments should ideally have been undertaken with several strains from each geographic location. This is because genetic drift (Chapter 1) can lead to the development of differences among strains while they are cultured in the laboratory. As a consequence, it is difficult to be sure that strains accurately reflect the populations from where they originated.

Apart from these limitations, geographical comparisons do not provide much evidence on whether adaptive responses have occurred specifically in response to extremes. The desiccation stress is extreme, and conditions that are lethal for one population are not lethal for another, suggesting that genes controlling responses to extreme conditions could have been selected. However, it is also possible that genetic differences developed for another reason. For instance, genes controlling survival under desiccation may also control egg laying or some other trait not associated with fitness under extremes. This raises the general problem of interactions among traits discussed in the next chapter. We really need to show that genes specifically controlling responses to extremes are under selection, or that differences among populations only become evident once extremes are considered.

Perhaps the best evidence for adaptive responses to climatic extremes in quantitative traits comes from reciprocal transplant experiments involving plants, because fitness can be measured under natural conditions. There are many experiments demonstrating that plants perform relatively better in environments from where they originated. Differences between populations can be large, to the extent that alien populations may fail to survive and set seed whereas local populations produce abundant seed. An example of this type of study is provided by Schmidt & Levin (1985) who undertook transplant experiments with an annual plant, *Phlox drummondii*. Reciprocal transplants were undertaken with eight populations over two years. The main environmental variable differing between sites was rainfall. Survival of seedlings varied from 0 to 92%, while the number of seeds produced per plant varied from 0 to 81. In general, local populations performed relatively better than the mean value of the alien populations for both survival and seed production. This is evident from the survival data for one year presented in Figure 3.5. However, note that some alien populations performed better than the local population at some sites, a common observation in these types of studies.

One reason why some alien populations may perform quite well in such experiments is that populations are adapted to conditions not present in particular years. For instance, many plant populations exposed to intermittently stressful conditions exhibit a slow growth rate. This may be an advantage when nutrients and other resources are not abundant because plants are less likely to exhaust resources (Grime, 1979; Chapin, 1980). Plants that grow slowly will usually be at a disadvantage, because they produce fewer seeds and may be outcompeted for light by rapidly growing individuals. However, under rare stressful conditions, these plants can be at an advantage because they continue to survive and reproduce, whereas rapidly growing plants may die as they exhaust resources. Comparing the relative fitness of plants over a short period can therefore lead to incorrect conclusions. Populations may become adapted to infrequent stressful conditions as well as the average conditions they experience. It is not clear how often this occurs or whether organisms are more likely to adapt to stressful conditions rather than average conditions, because experiments normally address only average conditions. However, evidence

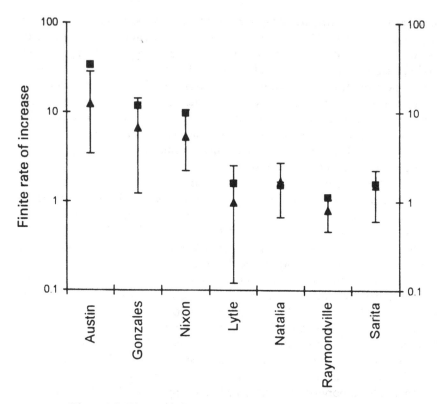

Figure 3.5. Fitness (finite rate of increase) of local and alien populations of *Phlox drummondii* tested at a series of sites in 1979. Squares represent values for the local population, and triangles represent the mean of the alien populations planted at the same site. Bars give the range of values for the alien populations. (Plotted from data in Schmidt & Levin, 1985).

discussed below indicates that the types of selection pressures experienced by populations can change markedly across years.

Enzyme polymorphisms and extremes

In the 1960s, protein electrophoresis was first used to examine the amount of genetic variability in populations. Because proteins are charged molecules, they migrate when a current is applied to a gel matrix. Small changes in the amino acid composition can change the total charge carried by a protein molecule, altering the way it migrates. Changes in DNA that result in a different amino acid strand can therefore be detected with this method.

By characterizing such variation at a number of loci coding for enzymes and other proteins, an indication of the overall level of allelic variation in a population can be obtained.

The fitness of allelic forms of enzymes (known as allozymes) has been related to sharp changes in environmental conditions. For instance, Johannesson, Johannesson & Lundgren (1995) considered allozyme variation in the enzyme aspartate aminotransferase from a marine snail (*Littorina saxatilis*) which occurs along the rocky shore. They found large differences in the frequency of an allele at this allozyme locus (Aat[120]). In the splash zone of the shore where snails are not immersed, the frequency of this allele was 90%; in contrast, its frequency was only 40% in the lower surf zone where immersion occurs, even though these sites were a few meters apart. A toxic bloom of a microflagellate in 1988 killed all the snails from the surf zone, but not the splash zone snails because these were not immersed. This led to recolonization of the surf zone by snails from the splash zone, and a concomitant increase in the frequency of Aat[120] in the surf zone. However, by 1992 the frequency of this allele in the surf zone had decreased to its original level. This rapid decline indicates strong selection against the Aat[120] allele in the surf zone. The strength of selection was determined by the computation of the relative fitness of different genotypes required to produce this change. This indicated that individuals homozygous for the Aat[120] allele had only 60% of the fitness of individuals with other genotypes. In other words, a genetic change at one enzyme locus may have been responsible for large differences in fitness between individuals, and the fitness of genotypes changed dramatically along sharp gradients in environmental factors. Selection is probably related to temperature and wetness as these factors differ markedly between the adjacent zones.

In another example, Vrijenhoek, Pfeifer & Wetherington (1992) considered variation at loci coding for four enzymes (the *Ldh-1*, *Idh -2*, *Pgd* and *CK-A* loci) in *Poeciliopsis monacha*, a stream-dwelling fish. Two alleles were distinguished at each locus. Fish were exposed to three stresses: hypoxia, cold and heat. For hypoxia and heat, fish carrying the common allele at each locus had relatively higher survival, and this effect was significant for several of the loci. In contrast, the alternate alleles appeared to be favoured under cold stress. It is not clear if selection acted directly on these loci or on areas of the chromosomes near these loci. Nevertheless, the data highlight that fitness differences can become evident under stressful conditions.

It has been suggested that fitness differences between allozymes will generally be more evident under stressful than favourable conditions. This follows from models (eg Kacser & Burns, 1981) relating the effects of enzyme activity on the rate of generation of end-products in a metabolic pathway, known as the 'flux' of a pathway. It is these products that influence the phenotype of an organism and ultimately its fitness. Normally, changes in enzyme activity have

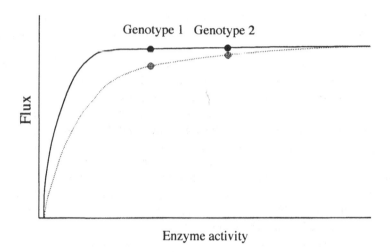

Figure 3.6. Asociation between flux through a biochemical pathway and changes in the concentration of an enzyme catalyzing a step in the pathway. The putative relationships are shown for favourable and stressful conditions. Differences between genotypes determining enzyme activity are more likely to become apparent in the stressful environment (From Hartl, Dijkhuizen & Dean, 1985).

minor effects on flux because the association between activity and flux is expected to follow the upper curve in Figure 3.6. Because an enzyme affects only one part of a biochemical pathway, activity changes associated with allozyme differences only have a small effect on flux. In stressful environments, the nature of this association may change. Researchers have argued that differences in the activity of one enzyme will then have some influence on overall flux through a pathway, as indicated in the lower curve in Figure 3.6, leading to a direct association between the effects of allozymes on activity and flux. However, the exact way environmental conditions change this curve remains unclear (Dean, 1994).

A connection has been made between enzyme activity and fitness under extreme conditions in a few cases. In higher plants, flooding reduces the availability of oxygen to roots, resulting in anaerobic stress and cessation of energy-producing metabolic pathways that require oxygen. This leads to a decrease in the uptake of nutrients and water, and consequently photosynthesis is reduced. However when flooding occurs, plants can continue to produce some metabolic energy by activating metabolic pathways involving fermentation. One of the enzymes involved in this anaerobic process is alcohol dehydrogenase (ADH). Chan & Burton (1992) related genetic variation in the activity of the ADH enzyme in white clover, *Trifolium repens*, to their environments. They collected plants from three sites with a different history of flooding within 20km of one another. Cuttings were taken from the sites and grown in controlled conditions. ADH activity in the roots was measured in the presence or absence of flooding. Measurements showed that ADH activity was induced to a greater extent in plants from flooded sites than in those from unflooded sites (Figure 3.7). Chan & Burton (1992) also showed that growth rate under flooded conditions was positively correlated with ADH activity, suggesting selection for increased activity in flooded areas. This example involves method 2 from Table 3.1 because genetic differences were found over relatively short distances. It also used method 8 because predictions about the involvement of ADH were made on the basis of prior knowledge about its role in fermentation. Note that Chan & Burton did not consider allozyme variation in ADH *per se*, but all genetic effects that influenced enzyme activity. Allozyme variation in ADH has been related to flooding in another plant, the soft bromegrass, *Bromus mollis* (Brown, Marshall & Munday, 1976).

As well as providing opportunities to study selection at specific loci, allozymes have also been extensively used as general measures of genetic variation in populations. By surveying variation at a number of enzyme loci, an overall indication of genetic variability in a population can be obtained. In addition, allozymes provide an indication of the overall levels of homozygosity/heterozygosity possessed by individuals. If individuals are characterized for variation at several enzyme loci, they may carry heterozygous genotypes at different numbers of these loci. Numerous attempts have been made to relate such overall heterozygosity levels to the fitness of organisms. These suggest that heterozygosity can often be positively associated with some measure of fitness when organisms are tested in a stressed environment, but not under favourable conditions. For instance, Scott & Koehn (1990) examined growth rate in the coot clam, *Mulinia lateralis*, under stressful and optimal salinity and temperature conditions. There was an association between heterozygosity and growth rate under stressful salinity or temperature conditions. This relationship was absent under optimal conditions or when both stresses were present. Similarly, Audo & Diehl (1995) considered the effects of soil moisture on growth rate in the earthworm, *Eisenia fetida*. They found no association between heterozygosity and growth rate under optimal moisture conditions. However, there was an association when moisture levels were lower; this relationship was stronger at moderate moisture levels that reduced growth by 50%, compared to low moisture levels that reduced levels by 75%.

A vertebrate example is provided by Teska, Smith & Novak (1990) who examined the association between heterozygosity and physiological traits in oldfield mice, *Peromyscus polionotus*. Mice were fed high-, medium- or low-quality diets, as determined by the amount of straw in the diet. Low-quality diets sharply reduced body weight. To examine patterns of food utilization, the efficiency whereby food was absorbed was examined under the different

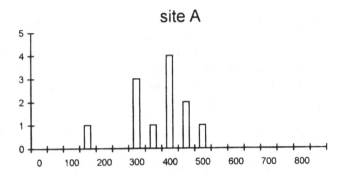

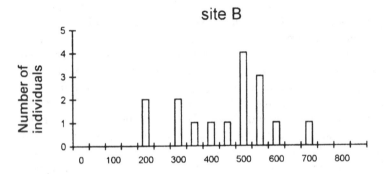

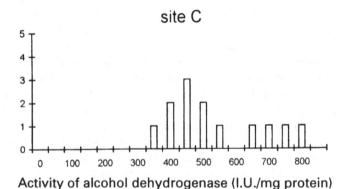

Activity of alcohol dehydrogenase (I.U./mg protein)

Figure 3.7. Distribution of activity of the enzyme alcohol
dehydrogenase in clones from clover plants originating from three sites.
Site A has not been flooded, whereas site C has been flooded more
frequently than site B. (After Chan & Burton, 1992).

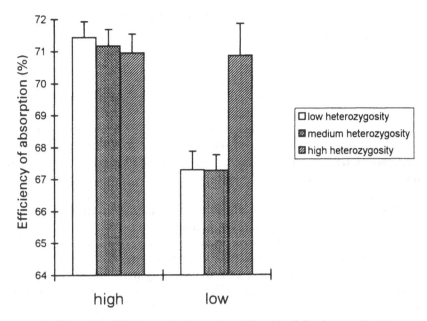

Figure 3.8. Efficiency of absorption of food (weight absorbed/weight eaten) for mice from three heterozygosity classes when exposed to high-quality or low-quality food. Individuals with a high heterozygosity level are at an advantage when food quality is low but not when it is high. (After Teska *et al.*, 1990).

conditions. Efficiency was expressed as the weight of food absorbed divided by the weight of food eaten. This measure indicated that heterozygosity did not influence absorption efficiencies when diet quality was high (Figure 3.8). However, while highly heterozygous individuals maintained their efficiency when fed low-quality diet, this was not the case for the other heterozygosity classes. Stressful diets therefore result in an association not evident under optimal conditions.

In summary, fitness variation at enzyme loci may be evident under stressful conditions but not under optimal conditions. Allozyme heterozygosity levels can be associated with growth rate when environmental conditions are unfavourable, although this may depend on the actual stress levels experienced. Presumably such associations reflect the combined effects of fitness differences between allozymes that become evident under stressful conditions. Allozyme heterozygosity may be related to the developmental stability of organisms (i.e. fitness), a concept we discussed in the previous chapter. There is increasing evidence that developmental stability is associated with heterozygosity under stressful conditions (Mitton, 1993).

Describing natural selection: quantitative approaches

Traditional techniques for investigating natural selection can indicate if selection has influenced trait values in populations. However, we often need more information to understand the selection process. If the distribution of a trait has changed during selection, this does not mean that the trait itself is under selection. The same genes can influence different traits, so changes may reflect selection on a second trait. For instance, selection might favour higher reproductive output by increasing the size of an organism. As a consequence, other traits influenced by size are also altered, such as fighting ability or development time. The way traits interact needs to be considered because selection may act on a suite of traits rather than individual traits. The ability of birds to withstand cold conditions will depend on their body fat, the efficiency of their metabolism as well as their overall size. All of these could be altered during a cold spell, but the extent to which they change will depend on any interactions between them. Genes increasing metabolic efficiency may decrease size or body fat, making it difficult to untangle the effects of selection by only looking at changes in individual traits.

Understanding how selection occurs also requires an understanding of how variation in a trait relates to variation in fitness. The simplest way is for increasing values of a trait to increase or decrease continuously with fitness (Figure 3.9(*a*)). In this case, selection will be 'directional' because one extreme of a trait's distribution is always favoured. Conversely, selection may be 'stabilising' when intermediate values of a trait are favoured (Figure 3.9(*b*)) or 'disruptive' if extreme values of a trait have relatively higher fitness (Figure 3.9(*c*)). Real curves may be more complicated than this. For instance, there may be a simple linear relationship between fitness and variation in a trait such as size. However, it is also possible that fitness differences are only evident when individuals are either very large or very small, leading to a non-linear relationship.

Several approaches for handling these complications have recently been developed as outlined in Brodie *et al.* (1995). A technique for understanding interactions among traits was developed by Lande & Arnold (1983), who proposed a quantitative method for identifying the strength of selection on each trait. Consider a population where a sample of n individuals has been scored for a continuous trait, x, and a measure of relative fitness, defined as w. The measure of fitness of individuals may involve simple categories, such as whether individuals live (fitness=1) or die (fitness=0), or may involve a continuous trait such as reproductive output or development time. To investigate if selection is operating and the type of selection that occurs, the association between x and w needs to be considered, by computing two variables, $x_1 = x$ and $x_2 = (x-\bar{x})^2$. If w is associated with x_1, there is evidence for directional selec-

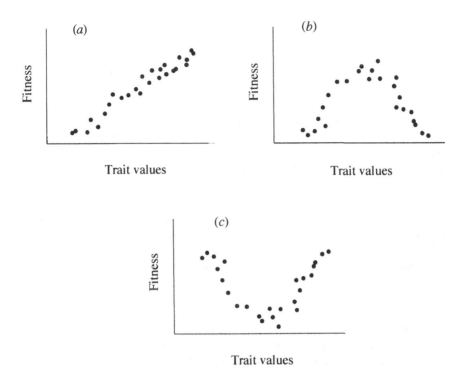

Figure 3.9. Three possible associations between variation in a trait and fitness. In (*a*), selection is directional because there is a direct association between increasing values of a trait and fitness. In (*b*), intermediate values have the highest fitness, leading to stabilizing or optimizing selection. Extremes have relatively higher fitness in (*c*), leading to disruptive selection.

tion as changes in a trait increase or decrease fitness. In contrast, x_2 will not be associated with w under directional selection because the extent to which a trait deviates from a mean value is, on average, not associated with fitness. However, x_2 will be associated with w if there is stabilising or disruptive selection, because in these situations large values of x_2 will be associated with high fitness values (disruptive selection) or small fitness values (stabilising selection).

The complete relationship between these variables and fitness can be described by the equation

$$w = a + b_1\, x_1 + b_2\, x_2,$$

where b_1 describes the effects of directional selection, b_2 the effects of stabilising selection, and a is a constant. For instance, if there is only directional selection, b_2 will equal zero, and the size of b_1 will reflect the intensity of selection.

To estimate the b_1 and b_2 coefficients, a technique known as 'linear regression' can be used to examine the relationship between the dependent values (w) and predictor variables (x_1, x_2). Because there is more than one predictor variable, researchers use a technique known as 'multiple regression', which allows the individual contribution of terms in the equation to fitness to be determined while holding other effects constant. It is possible to add additional terms to the regression equation. When there is directional selection, the association between a trait and fitness may not necessarily be linear, but may instead take the form of a curve as mentioned above. This can be detected by computing a new term, x^2, and using multiple regression to compute a new coefficient (b_3) for this term. The coefficients estimated in this way are known as partial regression coefficients.

Multiple regression can be used to consider the effects of more than one trait on fitness. For instance, assume that measurements are made on a second trait, defined by z, and that the association between fitness and the two traits is represented by the equation

$$w = 0.5 + (0.6)\, x_1 + (0.2)\, x_2 + (-0.3)z_1 + (0)z_2.$$

In this case, there is evidence for directional selection because variation in one trait (x) is positively correlated with fitness, while variation in the other trait (z) is negatively correlated. This means that once we control for variation in fitness due to x, increasing values of z are associated with decreasing fitness. There is also evidence for stabilising selection for x because the partial regression coefficient for x_2 is negative. The partial regression coefficients in this equation are known as selection gradients, because they describe the intensity of selection occuring on a trait. Selection gradients are referred to by the symbol β. They may be estimated for non-linear terms in the equation (the b_3 coefficient from above) as well as linear components.

This approach makes two main assumptions (Mitchell-Olds & Shaw, 1987). First, characters included in the analysis must completely describe the relationship between phenotype and fitness. If there are missing traits and these are correlated with traits being considered, the regression approach can lead to incorrect conclusions about the way selection influences traits. For instance, assume that the desiccation resistance of an insect was determined by its metabolic rate, which in turn was correlated with body size. If only size and desiccation resistance were measured during a dry spell, it might appear that selection was acting on body size, even though (unmeasured) variation in metabolic rate is the real target of selection. Second, it is assumed that the characters affecting fitness do so independently, without interactions. This assumption may normally be violated because interactions between characters are common. For instance, in the case of desiccation stress, fitness may be influenced by the hydrocarbons that help to waterproof an insect's exoskeleton as well as an insect's metabolic rate. These traits are unlikely to be inde-

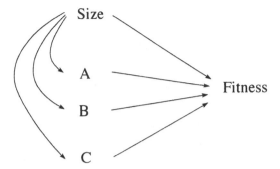

Figure 3.10. Path model for selection on morphological characters when general size influences fitness as well as three other traits (A–C) that also influence fitness.

pendent because metabolic rate depends on the movement of carbon dioxide and oxygen through the exoskeleton, which in turn is influenced by the effects of hydrocarbons on permeability.

One way of overcoming these assumptions is to undertake a 'path analysis' of selection (Crespi, 1990; Kingsolver & Schemske, 1991). This analysis depends on specifying the way traits interact with each other. An example of a path diagram describing a relationship between traits is given in Figure 3.10. In this case, one trait (general size) is assumed to influence the others (A, B, C). Size as well as the other three traits all influence fitness. This path diagram is likely to apply to the case where overall size influences other morphological traits which can in turn influence fitness. Under the multiple regression model described above, all of these variables would be expected to have a direct effect on fitness. Path analysis therefore can help to develop an accurate biological model of the way phenotypes are linked to fitness.

A major advantage of these approaches is that they lead to a quantitative description of the association between fitness and different traits. By estimating selection gradients, a fitness surface such as the one in Fig 3.11 can be constructed. This surface indicates that the effects of trait 1 and trait 2 on fitness lead to a dome shape. The combined effects of gestation age and birth mass on survival in humans approximates such a fitness surface (Schluter & Nychka, 1994). Fitness surfaces may be complicated and there are graphical techniques available that can help to describe them.

Describing natural selection: data

The multiple regression approach outlined above has been used to examine patterns of selection over time. Such studies highlight that the way

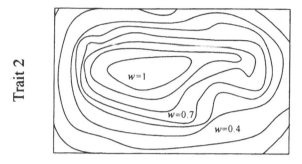

Trait 1

Figure 3.11. A two-dimensional representation of a dome shaped fitness surface. The fitness contours for changing values in two traits are given. Inner contours have the highest fitness (*w*).

selection operates on a trait can change across years. For instance, Kalisz (1986) considered selection on seed germination in the annual plant, *Collinsia verna*, over two years, by examining the effects of germination time on survival and reproduction. In the first year, there was directional selection because survival in winter and reproduction increased with early germination, while spring survival increased with late germination. There was no evidence for stabilising/disruptive selection. In the second year, these patterns changed dramatically. There was no evidence for directional selection on germination time, but there was disruptive selection for this trait. Different germination times are favoured at different times, making it difficult to predict long-term trends. Fluctuating patterns of selection are probably the norm in natural situations.

The multiple regression approach has been used to show that targets of selection can be difficult to identify from changes in individual traits. As mentioned in Chapter 1, drought altered the size of seeds available for a species of Darwin's finch, *Geospiza fortis*, in the Galápagos Islands. This led to a genetic change in morphology because large birds could utilize the hard seeds available during drought, and morphological traits in this species have a high heritability. To actually identify the targets of selection, Price *et al.* (1984) used the multiple regression approach of Lande & Arnold (1983). They first considered weight and wing measurements individually by computing selection differentials. They then computed selection gradients for these traits which took into account the interactions among traits. When individual traits were considered, it appeared that heavy birds with long and wide beaks were favoured during three drought years. However, a different picture emerged when all traits were considered; selection gradients were positive for weight and beak depth, but negative for beak width (Table 3.3). This indicates that drought conditions select for narrow beaks, presumably because these can

Table 3.3. *Selection gradients acting on Darwin's finches during three drought periods*

Standard errors are given in brackets.

	Drought period		
	1976–7	1979–80	1981–2
Weight	0.51±0.14	0.08±0.06	0.13±0.08
Beak length	0.17±0.18	−0.04±0.07	0.06±0.09
Beak depth	0.79±0.23	0.13±0.10	0.17±0.12
Beak width	−0.47±0.21	−0.14±0.09	−0.20±0.12

Source: From Price *et al.*, 1984.

handle the hard seeds. The reason why selection for a narrow beak is not evident from changes in the means of individual traits is that beak width is positively correlated with beak depth and overall size. In other words, changes in beak width were in the direction opposite to that imposed by natural selection because this trait was pulled along by its association with other traits that were also under selection.

Two studies on plants have considered how selection varies in different environments. The first (Bennington & McGraw, 1995) considered genetic differentiation in an annual plant, *Impatiens pallida,* from two sites. One of the sites is a floodplain, where plants grow in dense stands and reached 2 metres or more. The other site is a hillside where the plant density is much lower and plant height did not usually exceed 1 metre. The hillside site is associated with greater temperature fluctuations and reduced water availability, leading to a reduced lifespan and a reduced number of seeds. In reciprocal transplant experiments, populations performed relatively better in the sites from which they originated (Figure 3.12). This reflects differences in both survival and the number of seeds produced. To examine which traits were under selection, traits differing genetically between sites were examined. Plants originating from the hillside site tended to produce cleistogamous flowers (where fertilization occurs within an unopened flower) earlier regardless of the site where they were planted. The selection analysis revealed strong directional selection for this trait, because selection gradients are significant and negative. This suggests selection for rapid development in the hillside population to evade stressful conditions leading to early death of plants in this site. However, the same trend was not apparent for chasmogamous flowers (which open to allow cross pollination). There was also evidence for directional selection on some morphological traits. Plants from the floodplain population tended to have larger leaves and be taller than hillside plants. These characteristics may enhance

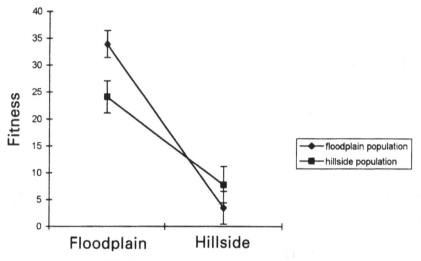

Figure 3.12. Overall fitness of populations of *Impatiens pallida* originating from two sites when grown at these sites. Both the floodplain population and hillside population are relatively fitter at their site of origin. Simplified from Bennington & McGraw (1995).

competitive ability at the floodplain site where plant density was relatively higher. In agreement with this conjecture, there was directional selection on leaf area at the floodplain site regardless of where the populations originated. The selection analysis therefore indicates different selection pressures acting at the two sites, which are consistent with genetic differences that have developed between the sites.

The second study used path analyses to examine interactions among traits and fitness. Jordan (1991) undertook transplants with two populations of the annual herb, *Diodia teres,* from an inland agricultural area and from a coastal sand dune habitat. Plants were initially grown in a controlled environment before progeny were transplanted to field sites. Differences between progeny were therefore likely to be genetic rather than being induced by environmental differences. A path analysis model was set up, based on the idea that the size of plants at one point in time determined plant size at a later time. In addition, size at each sampling time was assumed to influence seed production at maturity. A selection analysis of the model indicated that selection gradients were variable and changed at different times of the year. This meant that trait values favoured at one time were different to those favoured at other times. Nevertheless, there was a tendency for plant characteristics favoured at the coastal site to match those distinguishing the populations. For instance, the selection analysis indicated that short leaf length and heavy leaves were favoured at the coastal site. This is consistent with the relatively shorter and heavier leaves of coastal plants likely to be adaptive in dry sand dune environments where plants grow in exposed areas. However, traits selected at the inland site did not necessarily match differences between the populations. For instance, inland populations have thinner stems, but the analysis indicated selection at this site for thicker stems. Selection therefore varied markedly across environments and at different times, and did not always produce patterns that could be linked to population differences.

The studies discussed above are starting to provide insights into the way selection operates on traits. The effects of environmental conditions on the fitness of trait values appear to be variable in time and between environments. Selection gradients at a site may not always match the way populations from different sites have diverged. However, most studies utilizing quantitative techniques have focussed on the phenotypic level, and selection patterns at this level may not necessarily reflect those acting at the genetic level. As we discussed in Chapter 2, phenotypic variation can be due to the environment as well as genes, and genetic effects can interact with environmental conditions.

Genetic variation underlying evolutionary change: major and minor genes

There are two extreme views on the nature of genetic variation underlying adaptation to environmental changes. The first of these proposes that adaptation normally occurs via many genes each having a small effect on the trait under selection. These genes are known as 'polygenes' or 'minor' genes. The alternative view is that adaptation proceeds mainly by the selection of a few genes with a large effect on the trait under selection. These genes are known as 'major' genes. What is the evidence supporting these two extreme views of evolution in the context of adaptation to environmental extremes?

It is frequently assumed by researchers that genetic changes in a trait which is not under intense selection in one direction is largely attributable to minor genes. In this case, selection will favour a new optimum value not far from the initial optimum (see Figure 3.13(a)). Minor genes are more likely to be involved in such a change because it is thought that much of the genetic variance in a trait is associated with these genes. This is partly because mutations with large effects usually tend to have a low fitness. Because selection will act against major genes, they will be less common in populations than minor genes. Minor genes may also be favoured in this situation because many combinations of minor genes can achieve a new optimum value following a small environmental change.

In contrast, when adaptation requires a phenotypic change that is outside the normal phenotypic distribution of a population (Figure 3.13(b)), major genes are more likely to play a predominant role in the adaptation process

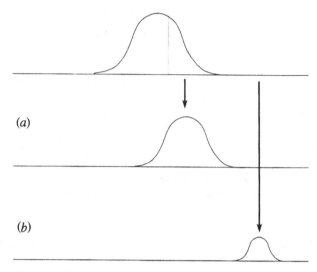

(a)

(b)

Figure 3.13. Response of a population to directional selection when (a) a new optimum is within the phenotypic distribution of the population and (b) a new optimum is well outside trait values encompassing most of the phenotypic distribution.

(Macnair, 1991). Under these conditions, selection is much more intense and only a few individuals with phenotypes well outside the mean of the population are expected to survive. Although combinations of minor genes may produce extreme values of a trait, these combinations will break up when individuals mate. For instance, consider the case where a major gene is important, and individuals homozygous for different alleles at a locus (say A_1 and A_2) mate to produce an F1 (genotype A_1A_2) that in turn mates with a similar genotype. The probability that an F2 individual has the A_1A_1 genotype will be $\frac{1}{4}$. Now if three unlinked loci control variation in a trait, the probability of crosses between individuals heterozygous at all three loci (say $A_1A_2B_1B_2C_1C_2$) producing a homozygous individual is only $(\frac{1}{4})^3$ or 1/64. When many loci are involved, there is only a low probability of recovering extreme genotypes. Major genes are therefore more likely to be important when selection is intense, and when a large phenotypic change is required to adapt to an environmental change.

We can use these arguments to explain the different types of genetic changes underlying the evolution of insecticide resistance in the field and the laboratory (Roush & McKenzie, 1987). Resistance in field populations of insects tends to be based on a single gene, whereas the response to selection for resistance in laboratory experiments tends to have a polygenic basis. In the laboratory, resistance is usually increased by selecting in the top 10–40% of survivors each generation, so the genes that contribute most to the phenotypic variance of a population will be involved in the selection response. These are likely to be minor genes. In contrast, field populations are much larger than laboratory populations. This makes it more likely that rare alleles with major phenotypic effects on resistance will occur. The dose of insecticide applied to a field population will be high because mortality levels approaching 100% are required for pest control. Individuals at the extreme end of the phenotypic distribution will therefore be selected, making it unlikely that minor genes will be involved. The genes contributing most to the phenotypic variance will therefore be unimportant if genetic changes occur in response to strong selection in one direction.

Similar considerations may account for the simple genetic bases that often underlie other adaptative changes to environmental changes imposed by human activities (Macnair, 1991). In rodents, selection for resistance to pesticides has involved changes at a single locus. In plants, the presence or absence of resistance to heavy metals associated with mine tailings often has a fairly simple genetic basis, although minor genes may influence the degree of resistance (Macnair, Smith & Cumbes, 1993). Adaptation to heavy metals has involved intense selection because only a few individuals from non-mine populations are usually able to survive on contaminated soils. Herbicide resistance in weeds may also be caused by major genes because selection imposed by chemical herbicides is likely to be as intense as that imposed by insecticides.

Such findings raise questions about the genetic basis of responses to other forms of environmental changes, in particular adaptation to climatic changes. Is natural selection imposed by climatic changes sufficiently intense to select only individuals at one extreme of a distribution? What is the genetic basis of phenotypic variation in natural populations for responses to climatic changes? Unfortunately, we do not have conclusive answers to these questions, even though the evidence considered above indicates that selection can be intense. Most data on traits involved in responses to climatic variables come from laboratory studies, but these studies may be biased in favour of minor genes for the reasons discussed above.

Some laboratory experiments have provided evidence that adaptation to high temperatures can involve major genes. For instance, Lenski & Bennett (1993) described experiments on bacteria to study the process of genetic adaptation to different temperatures. Replicate lines of *Escherichia coli* from a common ancestor were propagated for 2000 generations at 32 , 37 or 42°C. Several criteria indicated that 42°C was stressful for the ancestral clone while 32 and 37°C were not. For example, the ability of the ancestral clone to replicate was substantially reduced at 42°C, and this temperature was within 1°C of the temperature at which the clone became extinct. When lines of *E. coli* were propagated at the different temperatures, lines evolved much more rapidly in the thermally stressful environment (42°C) than at either of the lower temperatures. Despite this, adaptation to 42°C did not appreciably extend the highest temperatures bacteria could tolerate. None of the six lines tested could persist longer than the ancestral clone at 44°C. However, a population declining towards extinction would occasionally show a rapid recovery and thereafter maintain itself. This recovery was due to major genes that were referred to as 'lazarus' mutants. While these mutants did not always occur, their appearance in some populations suggests that large adaptive changes to high temperatures in *E. coli* may be achieved by occasional bouts of intense selection upon major genes.

Laboratory studies in *Drosophila* have also indicated that responses to selection for stress resistance can be associated with major genes. An advantage of *Drosophila* studies is that complicated crosses with special stocks allow genes affecting quantitative traits to be isolated to specific regions of the chromosome and ultimately the gene level. An outline of this technique can be found in Thoday (1979). Experiments using it have provided evidence that large differences in resistance can often be associated with specific regions on one or two chromosomes. For example, differences between strains of *D. melanogaster* resistant and sensitive to radiation and anoxia have been located to a few specific chromosomal regions, suggesting that major genes are involved in these traits (Parsons, 1973).

We can also obtain information about the genetic basis of variation in traits by carrying out a genetic analysis of differences between populations or closely related species that have diverged in response to an environmental change. This approach is being used more commonly with the advent of molecular markers that enables mapping of genes affecting quantitative traits. Such genes have become known as 'QTLs', which stands for 'quantitative trait loci'. In one approach for mapping QTLs, strains with different phenotypes are crossed, and the F1s are crossed to produce an F2 generation. The F2 individuals are scored for numerous marker loci and the quantitative trait. If different genotypes at a particular marker locus differ in their mean phenotype for the trait under investigation, this suggests that the marker locus is closely linked to a QTL. The approach depends on the availability of marker loci for a large part of the genome, and genetic variation detectable at the molecular level can provide such loci.

Recent applications of the QTL approach suggest that variation in quantitative traits may often be attributable to a small number of genes, in contrast to the classic view that quantitative traits are controlled by numerous genes with very small effects. For instance, in maize large morphological differences have been attributed to changes at only a few loci, while in *D. melanogaster*, variation in the number of bristles has been associated with a small number of loci controlling sensory organ development (Mackay, 1995).

The QTL approach has not yet been applied to traits involved in adaptation to environmental extremes. However, more traditional techniques have suggested that major genes may underlie stress evasion responses. For instance,

Rockey, Hainze & Scriber (1987) investigated pupal diapause in the eastern swallowtail butterfly, *Papilio glaucus*. In the northern subspecies adapted to a short growing season, diapause is obligatory and occurs each generation. In the southern subspecies, diapause is facultative and induced by short photoperiods. Crosses between these subspecies indicate that the incidence of diapause is associated with sex; when females from the southern species were crossed to males from the northern subspecies, almost all female F1 progeny diapaused, whereas there was a low incidence of diapause in F1 females from the reciprocal cross. Males from the reciprocal crosses did not differ for this trait. Unlike many other animals, female butterflies carry one X chromosome whereas males carry two X chromosomes. Because the female's X chromosome is inherited from the male parent, the difference in diapause induction between females from the reciprocal crosses indicates that one or more genes on the X chromosome control variation in diapause. The genetic basis of this trait may be simple because there is no evidence for the involvement of other chromosomes, although more crosses are required to confirm this. Data from crosses with other insects also provide some evidence that variation in diapause can have a simple genetic basis (Tauber, Tauber & Masaki, 1986).

A problem with comparisons of populations or species is that a large number of genes may appear to underlie differences between them even when adaptative changes are based on a few major genes. This is likely because genetic changes affecting a trait may occur *after* populations have diverged for this trait in response to an environmental change. Additional genetic changes may be completely unrelated to the original adaptive event. We can illustrate this with an example where an insect population has become adapted to dry conditions. If this population is also exposed to conditions that favour a low rate of egg production, there could be further changes in desiccation resistance because there are pleiotropic interactions between these traits (Rose *et al.*, 1992; Hoffmann & Parsons, 1993). When the populations are crossed, several genes may appear to control differences in resistance between the populations, even if adaptation to dry conditions only involves a few major genes.

One way of overcoming such problems is to consider adaptive variation within natural populations rather than between populations. An intriguing example is Smith's (1993) study of variation in bill size of the African finch *Pyrenestes*. This finch, which feeds on the seeds of sedges, is polymorphic for bill size. Birds with large bills feed more efficiently on a sedge species that produces hard seeds, whereas those with small bills feed more efficiently on soft seeds produced by another species of sedge. The fitness of these morphs will therefore depend on the relative abundance of the different types of sedges. Smith (1993) carried out crosses between birds with large and small bill morphs. Parents in these crosses tended to produce progeny with large or small bills but not with intermediate bill sizes (Figure 3.14). In addition, the relative number of large and small-billed progeny often fell into simple ratios. A

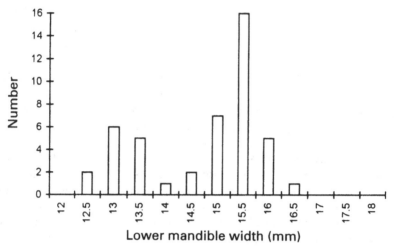

Figure 3.14. Distribution of bill sizes in *Pyrenestes* finches produced from crosses between individuals with large and small bills. (From Smith, 1993)

simple genetic basis is suggested when this occurs, and data were consistent with large size being determined by a single dominant allele. These findings suggest that adaptive changes in response to the abundance of different types of seeds could be mediated by a simple genetic change. However, more crosses are needed to verify the genetic basis of bill size.

This example represents an unusual situation in that bill size in *Pyrenestes* finches shows a bimodal distribution allowing birds to be classified as having either a large or small bill. In most species, variation in traits such as size will show a bell-shaped distribution rather than a bimodal one. However, a bell-shaped distribution tells us little about the genetic basis of a trait. We can only determine the nature of genes underlying traits involved in responses to environmental changes once crosses are undertaken between individuals with extreme phenotypes. Unfortunately, extreme phenotypes are most likely to come from widely separated populations exposed to different environmental conditions and with a history of selection for other traits.

Some general comments

Many of the examples discussed in this chapter indicate that natural selection has occured in direct response to environmental changes, but tell us little about the role of extremes in causing adaptive differences to arise between populations. In transplant experiments involving populations from different climatic regions, there is evidence that fitness differences between populations can be large. In many cases, organisms from alien populations fail to survive in local conditions. In other words, conditions that are extreme for one population can be tolerated by another population. Yet does this provide information about the role of extremes in adaptive divergence between populations?

We can illustrate this problem with two extreme scenarios about how adaptive divergence between two hypothetical plant populations might take place. Under the first scenario, plants growing in an environment experience conditions that change slowly. Assume that conditions are becoming drier. This leads to fitness differences between genotypes, but these are not particularly large because the drier conditions are not extreme. Genes increasing fitness in the drier environment are favoured, and these genes are likely to be those segregating at intermediate frequencies in a population. This process continues for many generations, eventually resulting in a population adapted to dry conditions. If genes that increase fitness in the new environment are associated with a decreased fitness under more moist conditions, this evolutionary process could result in large fitness differences among genotypes when populations are compared in reciprocal transplant experiments.

In the second scenario, environmental changes occur much more rapidly and conditions are extreme, at least intermittently. At the onset of dry conditions, most plants die, but a few manage to survive and produce offspring. These plants differ markedly in their responses to dry conditions compared to those from the original population. As a consequence, there has been a rapid evolutionary shift, which may be aided by the types of effects stress can have on genetic and phenotypic variability as described in the previous chapter. When the populations are eventually compared, there are large fitness differences, particularly if the genes having a large effect on fitness in dry conditions decrease fitness under optimal conditions.

Reciprocal transplant experiments will not distinguish between these scenarios. Only if the selection process was directly being observed could we unequivocally ascertain the role of extremes. However, it may be possible to distinguish between these alternatives by looking for indirect evidence. There are two ways of doing this. One way is to examine the genetic basis underlying adaptive responses to climatic extremes. We have seen that intense selection is likely to favour a simple genetic basis for traits, as evident in selective responses to stresses arising from human activities. If a few major genes are involved in an adaptive response, the inference is that a role for extremes becomes more likely.

The other way is to consider environmental effects on genetic variation. In the previous chapter, we established that extreme conditions can influence the expression of phenotypic variation and genetic variation in traits. If extremes persist, this can lead to selection on genotypes whose expression is not normally evident because of canalization. We could undertake experiments to see if this process has contributed to divergence in morphology and other traits between populations. By exposing populations from one location to the extreme conditions of another location, the effects of environmental conditions on the expression of phenotypic changes could be tested. By exposing an alien population to the same stress repeatedly over several generations, it may be possible to test if phenotypic changes can become assimilated. The equivalence of genes accounting for phenotypic changes in such experiments with genes contributing to differences between populations would need to be examined.

In contrast to geographic comparisons, the role of extremes can be determined more directly when populations occupy adjacent environments. In the case of heavy metal resistance, we know that changes occur over a short distance. Almost all plants from the original population do not survive in contaminated soils, indicating that conditions under which selection is occurring are extreme. Moreover, the genetic basis underlying the selection response appears to be relatively simple, reinforcing the likelihood of selection under extremes.

Natural selection in response to environmental changes occurring in time are also more easily interpretable in terms of extremes. In this case, we know that changes can be extreme because they often result in high levels of mortality and reproductive failure. Data from specific polymorphisms indicate that genetic variation is often under selection when extremes occur, whereas fitness differences between variants are not necessarily evident under favourable conditions. The techniques described above allow targets of selection to be identified. Studies such as those on morphology in Darwin's finches indicate that extreme conditions can impose intense selection on specific traits. Selection gradients indicate that the way selection acts on traits is not constant, changing from year to year depending on climatic conditions. Different extremes of traits can be favoured, or traits under stabilising selection in one year may be under directional selection in another year. This makes it most important to undertake long-term studies on populations to understand the way traits are likely to be altered by selection.

The arguments and examples considered in this chapter indicate that selection can occur under extreme conditions. However, we cannot state from the evidence reviewed that extremes account for most evolutionary changes in populations. One reason why many examples of natural selection revolve around extreme conditions is that the effects of selection may be more apparent at these times. Rare extreme conditions can cause an abrupt change in the mean of a trait, whereas weak selection imposed by predominantly favourable conditions may be difficult to detect, even though persistent favourable conditions will eventually influence the distribution of a trait. However, much remains to be done before the relative importance of extremes in evolution can be evaluated accurately. If the role of selection under extremes is to become clear, we might expect the types of patterns in the following list. Evidence is accumulating for some of these patterns.

(1)   When extreme conditions occur, and populations do not become extinct, directional selection changes the means of traits. The same traits may show stasis at most other times, either because traits are under stabilising selection or because selection is relatively weak.

(2)   When populations encounter unusual conditions, the populations are partly adapted to these conditions because intense selection in response to these conditions has previously occurred.

(3)   Adaptive responses occur under novel environmental changes that are extreme for organisms. Many environmental changes associated with human activities fall into this category.

(4)   The genetic basis of many adaptive changes is relatively simple because extremes have been important in adaptive divergence.

(5)   Selection on specific genetic polymorphisms is more easily detectable under extreme conditions than favourable conditions, because fitness differences are expressed at extremes.

Summary

Natural selection can be detected with a number of methods, and several are applicable to adaptation under extreme conditions. These include associations of variable traits and morphs with environmental factors at the geographical and local levels, and tracking changes in traits and morphs during stressful conditions.

Many of the best case studies of natural selection involve responses to stresses arising from human activities. These include responses to heavy metals, air pollution, insecticides and other stresses. Responses can be indirect and mediated via biotic factors, as in the case of industrial melanism. The genetic basis of traits involved in these responses tends to be simple.

Fitness differences among enzyme morphs tend to be larger under stressful conditions. This is reflected in associations between heterozygosity and fitness, which are more likely under unfavourable conditions.

Examples of natural selection due to stressful climatic changes include selection on the abnormal abdomen phenotype in *Drosophila* and shell banding in *Cepaea*. Comparison of populations for stress resistance are often difficult to relate to selective processes in populations. Reciprocal transplant experiments in plants provide conclusive evidence of genetic changes resulting in marked differences in resistance to extremes.

Regression methods are increasingly being used to identify the targets of natural selection and to describe the effects of selection on traits. Simple changes in the means of traits under selection may be inadequate to indicate selection targets. The types and intensity of selection acting on traits can change markedly across years.

Major genes can be important in adaptation to extreme conditions. In *E. coli*, genetic adaptation to extreme temperatures may involve genes with major effects. The intensity of selection can determine the nature of loci involved in responses to selection. Although major genes may be more important in adaptive responses than previously believed, more well-designed studies are needed to settle this issue. The role of extreme conditions in adaptive changes is unclear. However, a simple genetic basis for adaptive changes and the presence of environmentally-induced phenotypic changes may provide a starting point for isolating such effects.

CHAPTER 4

Limits to adaptation

From the previous chapter, it is clear that populations have often adapted to environmental changes, and that there is genetic variation in natural populations for quantitative traits likely to be involved in this process. The presence of variation suggests that populations often have the potential to evolve and adapt when they encounter environmental extremes. However, this potential appears limited because extinctions of populations and species are common.

In this chapter, we consider factors that can limit the ability of a species to counter environmental extremes. We will focus on costs involved in adapting to an environmental change by examining what happens to the stress response curve after selection has occurred. If costs are widespread, then adaptation to one set of conditions could reduce a population's fitness in another set of conditions. We also consider evolutionary changes at species boundaries. Boundaries often occur because populations have not been able to adapt to environmental conditions existing beyond them. We will emphasize an energetic framework in looking at boundaries and consider energetic costs arising from environmental stresses at boundaries. Finally, we briefly look at human adaptation to physical extremes of the environment. Although human populations can modify their environment to a large extent, there is some evidence that they have evolved in response to physical conditions of the environment that limit human distributions.

Changing the stress response curve

Costs associated with environmental adaptation are often viewed as changes in the stress response curve discussed in Chapter 1. If a two-tailed curve is considered as in Figure 4.1, a number of possibilities arise after a population has adapted to new conditions along an environmental gradient.

(a) An organism has evolved and performance in its original environment is not altered. The organism is therefore able to maintain a high level of fitness across a wider range of environments than before. We can talk of an organism having altered its plasticity level in this situation. As mentioned in Chapter 2, plasticity is defined as the extent to which the phenotype associated with a

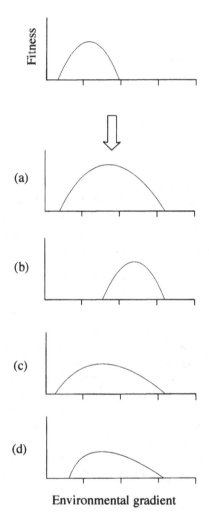

Environmental gradient

Figure 4.1. Possible changes in fitness along an environmental gradient after an organism has adapted to conditions towards the right of the gradient. The following changes may take place: (*a*) adaptation with little change in fitness in the original environment and a change in plasticity; (*b*) response curve moves to the right as adaptation is associated with decreased fitness in the original environment; (*c*) adaptation associated with a decrease in fitness under favourable conditions and a change in plasticity; (*d*) evolution involves both a shift in the response curve and a change in plasticity.

particular genotype can be altered by the environment. In this case, the fitness phenotype is less plastic after evolution has occurred because the same level of fitness is maintained over a wider range of environmental conditions.

(*b*) The breadth of the response curve has not been altered by selection but the curve has moved to the right. Here, increased performance at one environmental extreme is accompanied by a decrease in performance at the other extreme. We can therefore talk in terms of a 'tradeoff' having occurred between different parts of the environmental gradient. Overall, individuals have not become more or less plastic in their fitness, at least when the entire environmental gradient is considered.

(*c*+*d*) In these two situations, increased performance in a stressful environment is associated with decreased performance in an optimal environment. In 4.1(*c*), selection does not alter fitness at the opposite extreme, but there is a tradeoff between the two stressful environments and the optimum environment because the maximum fitness that can be attained is lower than before. The situation in 4.1(*d*) is similar except that a second tradeoff is evident between fitness at the two extremes along the environmental gradient.

What is the evidence that these types of tradeoffs and plastic changes actually occur, and moreover that they restrict evolutionary responses? From a biochemical and physiological perspective, tradeoffs between performance in different environments would seem likely. Four arguments have been proposed:

(1) Comparisons of the same enzyme from related animal species often suggest that enzymes perform differently at the same temperature, but function best at the temperature experienced by a species. An example (Hochachka & Somero, 1984) is provided by the enzyme lactate dehydrogenase from species of barracuda fish living in habitats with different temperatures. When three species are compared, their efficiency is almost identical at the temperatures they normally experience in their habitats (18, 23 or 26°C) despite differing by as much as 70% when tested at 25°C. However, it is usually difficult to relate differences in enzyme efficiencies directly to fitness.

(2) Traits that specifically increase survival under periods of environmental stress may decrease fitness when conditions are favourable. For example, increased resistance to a range of stresses in animals seems to be associated with a low basal metabolic rate, particularly in response to reduced food availability (Merkt & Taylor, 1994). However, a low rate may decrease fitness under optimal conditions because it can decrease mating success and the number of progeny females produce (Hoffmann & Parsons, 1991).

(3) Structural costs may be associated with the machinery and processes animals have to maintain in order to be resistant to stresses. Thick protective coats or fat reserves can reduce the manoeuvrability of animals and increase their susceptibility to predators or decrease their sexual activity. Proteins needed to detoxify chemicals may interfere with a cell's normal functioning.

Resistance or evasion mechanisms may take up valuable space within a body cavity, decreasing space available for storage of eggs and reproductive organs. (4) Organisms have a finite amount of resources available to them, and any shift in allocation to one component of metabolism is necessarily at the expense of other components (Sibly & Calow, 1989). Countering environmental stresses often requires energy and other resources. Many animals need energy to produce protective coats that keep cold environments or toxins out. They may also accumulate large amounts of fat to increase survival during periods of food shortage or drought. Any deviations from optimal conditions requires an increased expenditure of energy. For instance, Zotin (1990) showed that the minimum energy expended by the three ontogenetic stages (embryos, larvae, pupae) of *D. melanogaster* occurs at around 23°C. Metabolic efficiency, measured from the amount of energy required from a female to produce an average egg at various temperatures, is higher in the 22–25°C range than at 18°C and 28°C (Figure 4.2). Maximum reproductive potential therefore occurs in the intermediate temperature range that incorporates the region of minimum metabolic cost.

Have these factors resulted in the types of tradeoffs envisaged in Figure 4.1? If so, what are the underlying physiological variables, and to what extent do tradeoffs constrain evolutionary change? We can examine these questions by looking at several studies of tradeoffs using a variety of approaches.

Studies of tradeoffs

1. *Stress resistance and life history traits in* Drosophila

A powerful technique for investigating tradeoffs is to select artificially for increased performance under one set of conditions, and examine performance under different conditions after several generations of selection. Tradeoffs are suggested when there is a decrease in performance under conditions not being used to carry out selection.

Several selection experiments with *Drosophila* have indicated a tradeoff between the ability of flies to resist stressful conditions and their fitness under optimal conditions (Hoffmann & Parsons, 1991). Increased resistance to a number of stresses has been selected including resistance to starvation, toxic ethanol levels, and desiccation. An example of this approach is our desiccation selection experiment with *Drosophila melanogaster* (Hoffmann & Parsons, 1989). Females were exposed to dry conditions until most (85%) of them had died. Survivors were transferred to moist conditions and the progeny of these females were selected in the same manner. After several generations of selection, the desiccation resistance of the lines increased substantially.

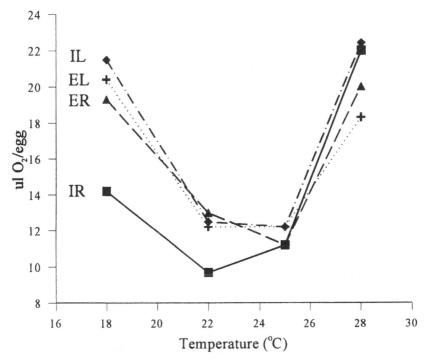

Figure 4.2. Effect of temperature on the metabolic efficiency of four strains of *D. melanogaster*. Efficiency is measured as the amount of energy required from a female to produce an egg. (From Arking *et al.* 1988).

To test if this change was associated with a decrease in fitness under favourable conditions, flies from selected and unselected lines were compared for a number of traits related to fitness (Table 4.1). Females from the selected lines laid fewer eggs early in their life when they were provided with ample food. Selected females therefore produce fewer progeny under optimal feeding conditions, suggesting a tradeoff between fitness in stressful and optimal environments. A tradeoff between egg laying (fecundity) and stress resistance has also been identified from an experiment involving selection on egg laying rather than resistance. Service, Hutchinson & Rose (1988) selected *D. melanogaster* females that produced a large number of eggs early in their life. They found that starvation resistance in the selected lines decreased. Genes increasing early fecundity therefore reduced starvation resistance. This represents an example of increased performance in an adverse environment lowering performance under optimal conditions.

Some progress has been made in understanding the physiological basis of the tradeoff between stress resistance and early fecundity. Three factors that

Table 4.1. *A comparison of control (desiccation-sensitive) and desiccation-resistant strains of* D. melanogaster *generated by artificial selection*

Trait	Control		Selected (resistant)
metabolic rate	high		low
starvation	sensitive		resistant
heat	sensitive		resistant
intense γ-radiation	sensitive		resistant
toxic ethanol exposure	sensitive		resistant
toxic acetic acid exposure	sensitive		resistant
cold shock		no difference	
acetone (toxic)		no difference	
ether (toxic)		no difference	
lipid content		no difference	
early fecundity	high		low
early behavioural activity	high		low
development time		no difference	
mating success		no difference	
male longevity	short		long

Source: Compiled from data in Hoffmann & Parsons (1989, 1993 and others).

have been implicated are the metabolic rate of the flies and the amount of lipid and glycogen reserves that they carry. Flies from lines that are resistant to a range of stresses often have lower metabolic rates, as in the case of lines we selected for desiccation resistance (Table 4.1). This can increase starvation resistance by reducing food requirements. A lower metabolic rate can also increase desiccation resistance by reducing the need for gaseous exchange through an insect's spiracles, because spiracle opening is an important source of water loss in insects. Glycogen and lipid reserves can increase starvation resistance by providing food reserves, and they may also play a protective role by reducing the amount of water lost from flies (Rose *et al.*, 1992).

The tradeoff between early reproduction and survival in stressful environments has also been related to longevity. Strains of *D. melanogaster* that are more stress resistant and have lower early fecundity tend to live longer, suggesting that genes influencing stress resistance also influence survival under optimal conditions (Service *et al.*, 1988; Rose, 1991; Table 4.1). Longevity has been associated with lipid/glycogen levels and metabolic rate. An enormous body of data on *Drosophila* has demonstrated that longevity varies inversely with specific metabolic rate. In particular, environmental manipulations that increase metabolic rate tend to decrease longevity (Miquel *et al.*, 1976). These manipulations include changes in temperature, sexual activity and flight ability.

An association between survival under adverse conditions, longevity and metabolic rate is evident in comparisons of mutant strains as well as in selected strains. Mutations are usually isolated by looking for drastic phenotypic changes. For example, 'shaker' mutants of *D. melanogaster* are characterized by high behavioural activity levels as implied by their name. These mutants also have increased metabolic rates and reduced longevity and courtship success when compared to normal flies. Furthermore, these and other high metabolic rate mutants and strains are sensitive to environmental stresses including high temperature, and an unsaturated aldehyde, acrolein. One of the 'shaker' mutants, known as *hyperkinetic[1]*, is sensitive to Los Angeles smog, especially in combination with temperature stress. High sensitivity to environmental stresses follows from the cost of the substantial drain of metabolic energy measured by oxygen consumption that is characteristic of these strains (Parsons, 1992b). Mutants influencing longevity have also been isolated in other organisms. In the nematode, *Caenorhabditis elegans*, a mutation in the *age-1* gene causes an increase in lifespan of 65%; in addition, this mutation is associated with increased resistance to high temperatures, suggesting an association between stress resistance and longevity (Lithgow *et al.*, 1995).

2. Evolution of high temperature resistance in E. coli

Another way we can demonstrate tradeoffs is to allow organisms to 'evolve' under one set of conditions and then examine their performance under a different set of conditions. The bacterium, *E. coli*, has been used in a number of experiments involving laboratory evolution. Bacteria have a number of advantages for such studies. They have a rapid generation time, so that evolutionary changes can be followed over hundreds of generations. Their laboratory environments can be rigorously controlled by precisely defining culture conditions. In addition, evolutionary changes can be monitored accurately because stocks can be kept in an inactive state, enabling a direct comparison between evolved lines and ancestral stocks after an experiment has been completed.

Experiments on laboratory evolution for thermal stress in *E. coli* mentioned in Chapter 3 are summarized in Lenski & Bennett (1993). These researchers started a series of populations with *E. coli* derived from the same clone. The clone they used did not have plasmids or viruses that can mediate recombination in bacteria. Because recombination was absent, any genetic changes that took place in the lines were the consequence of mutation. The ancestral population from which the clone had been derived was adapted to growing at a constant temperature of 37°C, and populations were set up at this temperature or at 42°C. The higher temperature was stressful for this clone.

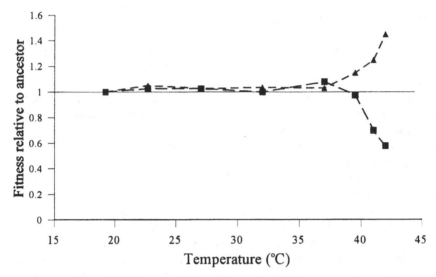

Figure 4.3. Changes in fitness in strains of the bacterium *E. coli* following adaptation to 37°C (squares) or 42°C (triangles). Fitness is expressed relative to an ancestral strain from which the evolved strains were derived. (From Lenski & Bennett, 1993).

Populations were allowed to evolve for 2000 generations, and then tested for adaptation to the different temperature conditions. When the lines exposed to 37°C and 42°C were compared, Lenski & Bennett (1993) found that all lines had evolved to be fitter than the ancestral line at the temperatures under which they were selected (Figure 4.3). However, there was no evidence for tradeoffs involving either the lower temperature extreme or the optimal temperature, because the 42°C lines performed as well as the ancestral clone and the 37°C line at these temperatures. Selection at the high temperature extreme therefore seems to have increased the breadth of the resistance curve without tradeoffs, as in Figure 4.1(*b*).

However, the 37°C lines showed a different trend, because these exhibited a decrease in competitive ability at higher temperatures relative to the ancestral population. This suggests that continued evolution at an optimal temperature may eventually result in a tradeoff between optimal conditions and high temperature extremes.

3. Haemoglobin in deer mice

We can also obtain evidence for tradeoffs at the level of a single gene. If one allele at a locus is favoured in an environment and an alternative allele is favoured in a different environment, then variation at the locus can contribute to a tradeoff between fitness in the two environments.

If variation in the DNA sequence at a locus is to have an effect on fitness, it should influence the types of gene products produced by the locus. These gene products should, in turn, have an effect on the fitness of an organism. While linking variation at a particular locus to fitness is a difficult task as discussed in Chapter 3, this has been achieved in several cases, including fitness differences related to climatic conditions.

As an example, we consider a study by Chappell & Snyder (1984) on variation in the haemoglobin chain of the adult deer mouse, *Peromyscus maniculatus*. This species is found at a wide range of altitudes in North America, spanning from sea level to above 4300 m. Haemoglobin is responsible for carrying oxygen around the blood. The haemoglobin molecule is complex, consisting of two types of chains of amino acids which are known as alpha and beta chains. In deer mice, alpha chains are coded by two loci that are tightly linked, known as *Hba* and *Hbc*, and there is genetic variation at both loci. Because the loci are tightly linked, there is little recombination between them, meaning that they are inherited together. An allele at one of these loci therefore tends to be associated with an allele at the other locus and these allele combinations are coinherited. Combinations of alleles that are normally associated with one another because of linkage are known as 'haplotypes'.

In deer mice, allelic variation in alpha chains produced by the *Hba* locus can be divided into two groups, a^0 and a^1. Variation at the *Hbc* locus can also be divided into two groups known as c^0 and c^1. The non-random association between these loci means that alleles with the same superscript almost always tend to occur together (i.e. a^0 with c^0 and a^1 with c^1). Chappell & Snyder (1984) showed an association between haplotype frequencies and the fitness of deer mice at different altitudes. The a^0c^0 haplotype increased in frequency with altitude in various populations. At high altitudes, the partial pressure of oxygen is much less than at sea level, restricting the availability of oxygen for metabolism. The different haplotype frequencies suggest that one of them is better at countering this lower oxygen pressure. Chappell & Snyder (1984) tested deer mice that were homozygous for the two haplotypes (a^1c^1/a^1c^1 and a^0a^0/c^0c^0) as well as heterozygous individuals, and they examined the affinity of haemoglobin from each strain for blood oxygen. They found that the a^1c^1/a^1c^1 deer mice had the highest oxygen affinity, while the heterozygote had intermediate values. In addition, they found that the a^1c^1/a^1c^1 homozygotes had the highest recordings for maximum oxygen consumption at low altitudes, whereas the a^0a^0/c^0c^0 homozygotes had the highest recordings at high altitudes.

The authors suggested that this physiological difference is directly attributable to genetic variation in the alpha chains. Theoretical models indicate that, in order to maintain optimal oxygen transport and maximal metabolic rate, oxygen affinity should increase with altitude until intermediate altitudes are reached, but affinity should decrease and fall at very high altitudes to values below those at sea level. The a^0c^0 haplotype is therefore likely to be favoured at high altitudes. This is consistent with the higher metabolic rate expressed by this haplotype at high altitudes and with altitudinal changes in haplotype frequencies. Variation at loci coding for alpha chains therefore seems to result in a genetic tradeoff between fitness in high and low altitude environments. The genotype with a higher fitness peak in one environment is likely to have a lower fitness peak in a different environment. Although fitness curves for the genotypes have not been measured over a range of environmental conditions, this situation probably corresponds to a shifting response curve as in Figure 4.1(*b*).

4. Resistance to chemical stresses

As discussed in Chapter 3, resistance to chemical stresses is often associated with single genes. Many studies have demonstrated that these genes are associated with fitness costs in the absence of the chemical. In particular, comparisons of strains of insects that are resistant or sensitive to a pesticide often indicate that resistant strains have lower fitness than susceptibles when the pesticide is absent, although costs have not been detected in all comparisons (Roush & McKenzie, 1987), particularly when environmental conditions are favourable.

There are also well-documented cases of costs associated with pesticide resistance in vertebrate pests. An example is the resistance of rats to a poison known as warfarin. In Britain, this poison has been used in the control of rats since the early 1950s. Resistance to warfarin was first detected in the late 1950s. Within a population, resistance is due to a single gene, although different resistance genes seem to be involved in different populations.

Costs are associated with warfarin resistance because resistant genotypes have an increased requirement for vitamin K (MacNicoll, 1988). This vitamin is involved in blood clotting. Individuals which are homozygous for the resistant allele suffer from disorders associated with clotting when vitamin K is absent. This in turn influences the growth rate and survival of rats (Smith, Townsend & Smith, 1991). As a result, resistant individuals have a lower fitness under field conditions. Costs associated with resistance to warfarin are therefore likely to account for declines in the resistance allele that are often detected when rat populations are no longer exposed to the rodenticide.

Warfarin resistance provides an example of fitness costs detectable under field conditions. Increased survival under adverse conditions is associated with decreased survival in more favourable environments where the pesticide is absent, following Figure 4.1 (*c*)/(*d*).

5. Darwin's finches and body size

In the previous two examples, tradeoffs were detected under field conditions by focussing on a single gene. Tradeoffs in the field can also be detected by examining variation in a particular trait with a known genetic basis, as in the case of selection on the body size in the Darwin's finch, *Geospiza fortis*, discussed previously in Chapters 1 and 3. Tradeoffs were detected by a detailed study of body size in one population of *G. fortis* during environmental changes.

During drought conditions, many of the adult finches die because of a diminishing supply of seeds. Price *et al.* (1984) showed that during each period, there was selection for birds with narrow beaks and a large body size. These birds were able to crack large and hard seeds that form the predominant food supply during drought periods. Small birds died because they could not process such seeds, and mortality levels up to 85% of the adult population were recorded.

In contrast, small birds are favoured under wet conditions. Gibbs & Grant (1987) found that small birds had higher survival rates than large birds in an exceptionally wet year. Small birds are probably favoured because the food supply consists mainly of small soft seeds that are more efficiently handled by small individuals.

Variation in body size and other morphological traits in Darwin's finches are largely determined by genetic factors. The high heritability for such traits may partly reflect the effects of hybridization between species that are not completely isolated reproductively (Grant, 1986), a process that is likely to continually introduce new genes into the population. Because the heritability of these traits is high, climatic changes lead to oscillating selection for genes controlling body size. Genes increasing size are favoured in dry environments, whereas those decreasing size are selected under the opposite extreme. Body size can therefore accomodate a tradeoff between dry and wet environments in *G. fortis*.

6. Flightless insects

Insects often migrate to evade stressful environmental conditions. In some groups such as aphids, grasshoppers, waterstriders and planthoppers, the same species may consist of individuals with functional wings that can migrate, and those that cannot. The latter may be wingless, or they may have reduced wings or flight muscles that prevent flight.

It is generally believed that flightless forms of insects evolve because of costs involved in maintaining wings and flight muscles. Many studies have demonstrated that flightless forms produce more offspring and start to reproduce earlier than migratory forms (Rankin & Burchsted, 1992). Costs are expected to be associated with flight ability. Metabolic costs may arise because of the energy allocated to the production of wings and flight muscles. In addition, the space occupied by flight muscles may impose a structural cost, such as by reducing the area a female has available for the storage of her eggs.

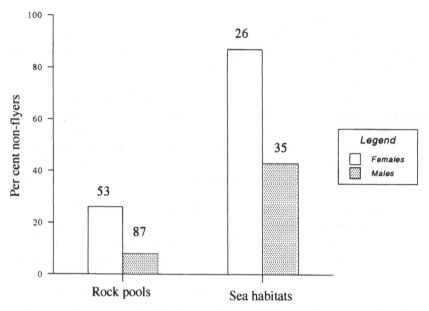

Figure 4.4. Incidence of waterstriders of the species, *Gerris thoracicus*, that cannot fly. Waterstriders were collected from two types of habitats in southwestern Finland. (Redrawn from Kaitala, 1988).

For example, populations of the waterstrider, *Gerris thoracicus*, consist of two morphs. One of these maintains its flight muscles (and therefore its ability to fly) throughout its reproductive period, whereas the other morph breaks down these muscles, preventing flight. Kaitala (1988) compared the fitness of these morphs in the laboratory to test for costs associated with flight ability. When food was abundant, females maintaining flight muscles laid 60–100% fewer eggs than females without these muscles. When food was scarce, flightless females survived longer than those with flight muscles, although there was no difference in survival among the males. Maintaining flight muscles is therefore associated with a cost in terms of reproductive output and survival.

Such costs suggest that the absence of flight may be selected in habitats that are stable, whereas flight ability may be selected in unstable habitats where organisms frequently encounter periods of environmental stress. In *Gerris thoracicus* from Finland, morphs that are able to fly are more common in temporary rock pools where a supply of food is likely to disappear when ponds dry up (Figure 4.4). In contrast, flightless morphs are more common in sea bays where a permanent food supply exists. Flight ability can therefore form the basis of a tradeoff between environments. This represents an example of Figure 4.1(*c*) where the fitness of phenotypes favoured during stressful conditions is reduced under more favourable (stable) conditions.

Complications

The above examples illustrate that tradeoffs can occur between environments, but there are two complicating factors that we need to consider in these types of studies. The first of these is acclimation, which can complicate the detection of tradeoffs. The second is 'genetic coadaptation', a process that can modify genetic costs associated with mechanisms of overcoming environmental stresses.

1. Acclimation

The performance curves in Figure 4.1 are simplistic in the sense that they ignore the potential for acclimation when an environmental change takes place. Acclimation can be regarded as an additional source of plasticity. As discussed in Chapter 1, many organisms are able to increase their resistance to extreme conditions by becoming acclimated. These acclimation responses may occur in response to non-lethal levels of an environmental variable, or in response to a variable unrelated to the stress itself such as a shift in daylength. Once an organism has been exposed to conditions leading to acclimation, its response curve can change markedly.

Acclimation introduces another dimension to the question of tradeoffs. Instead of simply asking whether tradeoffs occur between two sets of conditions, we need to consider whether tradeoffs occur when a population adapts to conditions via changes in its ability to acclimate. Tradeoffs that may arise because of acclimation are shown in Figure 4.5, which considers the fitness of an adapted population under conditions that do or do not result in acclimation. If adaptation has not involved acclimation, then the response curve will move to the right regardless of the conditions a population experiences (Figure 4.5(*a*)). However, if a population has evolved the ability to become acclimated in the period preceding the onset of stressful conditions, it will have a broader fitness curve if permitted to undergo an acclimation response as indicated in Figure 4.5(*b*). There may be a tradeoff associated with an evolved ability to acclimate, in which case the adapted population may show decreased fitness when an acclimation response is not triggered (Figure 4.5(*c*)).

Costs are likely to be associated with an increased ability to acclimate although direct evidence for costs is limited (Huey & Berrigan, 1996). In order to make an acclimation response, an organism needs to detect a stress before it occurs. It also needs to produce physiological and biochemical changes comprising the acclimation response. The machinery needed for an acclimation response may have energetic costs or other costs because of interference with other metabolic processes.

There is ample evidence that acclimation responses influence fitness in non-stressful environments. An example is provided by the response of the house-

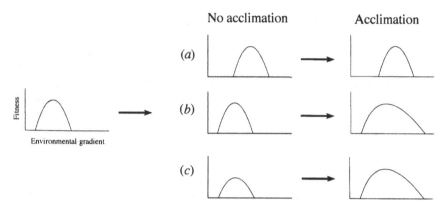

Figure 4.5. Possible involvement of acclimation when populations adapt to conditions towards the right of an environmental gradient: (*a*) adaptation involving a change in the fitness response curve without the evolution of an acclimation response; (*b*) adaptation because populations become acclimated to the new environmental conditions; (*c*) population changes in acclimation ability, but accompanied by a decrease in fitness in conditions when acclimation does not occur.

fly (*Musca domestica*) to a cold stress. Pupae of this species can attain a high level of cold resistance when they are briefly exposed to a non-lethal temperature. When Coulson & Bale (1992) compared females emerging from acclimated pupae to those emerging from pupae that had not been acclimated, they found that those in the former group lived five days shorter and also produced fewer eggs in the absence of a cold stress. As a consequence, the total number of eggs produced by females from acclimated pupae was only 55% of the numbers produced by non-acclimated control females. The physiological machinery induced by cold acclimation in *M. domestica* seems to have a large deleterious effect on the fitness of flies in favourable conditions. Similar fecundity costs have also been described in *Drosophila* for acclimation to increase heat resistance (Krebs & Loeschcke, 1994), although it is difficult to separate costs of acclimation from injury caused by the acclimating conditions themselves.

In such studies, we only have indirect evidence for a cost to acclimation. The reduction in fitness following acclimation suggests that individuals with and without an ability to undergo acclimation would also show differences in fitness. However, we should really compare genotypes differing in acclimation ability if we want to demonstrate tradeoffs. If tradeoffs exist, genotypes with a relatively greater ability to acclimate to a stress should have a lower relative fitness when the stress is absent.

Unfortunately, there is very little information on differences in the acclimation responses of genotypes. Most evidence for costs associated with acclimation comes from comparisons of different species, particularly in plants (Hoffmann & Parsons, 1991). For example, plant species that occur soon after the vegetation of a disturbed area has been removed are able to alter their rates of photosynthesis in response to light and shady conditions. This ability allows the plants to grow rapidly after a disturbance when high rates of photosynthesis in bright conditions are possible. Such species can also persist in the shady conditions that develop as new vegetation becomes established. However, they are eventually displaced by other species that show less flexibility in their rate of photosynthesis. Because the less flexible species are more successful under shady conditions, there may be a cost associated with photosynthetic flexibility.

2. Coadaptation

Costs associated with adaptation to new environmental conditions may be partly or completely overcome by the evolution of other compensatory changes. This is an example of a process known as 'coadaptation'.

We can illustrate this process with an example (McKenzie & Clarke, 1988). An insecticide known as Diazinon has been widely used in Australia to control populations of the sheep blowfly, *Lucilia cuprina*. Blowflies have evolved partial resistance to this insecticide, but Diazinon continues to be used because it is still partially effective. Resistance is associated with a single gene, so that three genotypes, RR (homozygous for the resistance allele), SS (homozygous for the sensitive allele) and SR (heterozygous) can be identified.

When experiments were carried out with blowflies from an area where the insecticide had been applied for some time, the fitness of the three genotypes in the absence of the insecticide was found to be similar. However, a different picture emerged when these genotypes were placed on a different genetic background. This was done by crossing heterozygous individuals to a laboratory stock that was susceptible to diazinon. Heterozygous progeny were also crossed to this stock, and this 'backcrossing' procedure was repeated for several generations. As a consequence of this process, the R gene was no longer associated with the genetic background of the field population, but instead became associated with the genetic background of the susceptible laboratory line. When the three genotypes were compared after backcrossing, the SS genotype was now fitter than the other two genotypes, suggesting that there were costs associated with the resistance gene.

McKenzie & Clarke (1988) showed that an evolutionary change had occurred at a second gene located within the same area of the chromosome as the R gene. When the resistance gene first arose, it was associated with a fitness cost. However, selection at a second locus led to an increase in the frequency of an allele that modified this cost and largely removed it. Such genes are therefore referred to as 'modifier' genes. In the case of diazinon resistance in

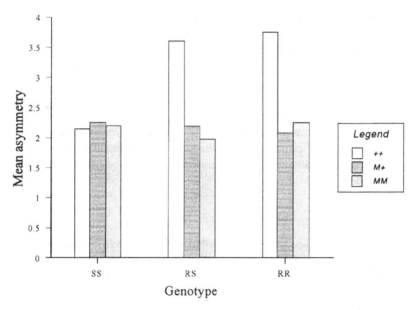

Figure 4.6. Fluctuating asymmetry in genotypes of the sheep blowfly, *Lucilia cuprina*, resistant and susceptible to the insecticide Diazinon. Asymmetry in resistant (RR) and heterozygous (RS) genotypes is influenced by genotypes at a second locus (M). When the ++ genotype is present at this second locus, asymmetry levels are markedly higher in resistant individuals. (From McKenzie & Clarke, 1988).

blowflies, the modifier had no influence on the fitness of SS homozygotes, and only improved the fitness of genotypes with the R allele. We can therefore talk about the modifier and R allele as being adapted to act together, or as being 'coadapted'. The effect of this modifier can be clearly seen by examining asymmetry in blowflies with the different genotypes (Figure 4.6). As discussed in Chapter 2, asymmetry can be used as an indicator of the fitness of an organism. The modifier does not influence the asymmetry of the SS flies, but causes a marked decrease in the asymmetry of the RS and RR genotypes.

The effects of coadaptation can be identified when stress resistance is controlled by a major gene as above, because backcrosses can be carried out to place the major gene in different genetic backgrounds. However, this is not feasible when resistance is controlled by several loci because resistance alleles at different loci would not be inherited together. For this reason, the role of coadaptation in minimising fitness costs associated with other stress responses is not known.

How common are tradeoffs between extreme and favourable conditions?

The above examples highlight the diversity of approaches that can be used to demonstrate tradeoffs between fitness in extreme environments and favorable conditions. However, they indicate little about the importance of tradeoffs. The general consensus among evolutionary biologists is that tradeoffs between environments are fairly uncommon. This is based on empirical data comparing the same genotypes across a range of environments. In general, correlations among environments are positive or zero rather than negative, arguing against tradeoffs (e.g., Andersson & Shaw, 1994; Windig, 1994)

There are two limitations inherent in most of these studies for answering questions about tradeoffs involving extremes. Firstly, the environmental conditions encompassed often span a fairly narrow range. For instance, temperature studies on *Drosophila* often involve experiments at temperatures ranging from 18°C to 25°C, while experiments on nutrition in mice often involve comparisons of diets that reduce growth rates by only 10 or 20%. These types of conditions do not encompass extremes, and cannot be used to address the question of tradeoffs involving extremes. Secondly, studies often focus on specific traits rather than the effect of a trait on fitness. High values for a trait may be selected in one environment, while low values may be selected in a different environment. Positive genetic correlations across environments based on trait values may therefore not provide much information on tradeoffs. We have already considered examples above where different trait values are favoured in different environments.

There are also limitations in using a correlation approach for detecting tradeoffs. In most studies, a range of genotypes is normally scored under different conditions, and correlations are carried out among genotypic scores. The genotypes may represent different strains, families or selection lines. As a consequence, all genotypes contribute to the correlations between environments or between traits. The correlation represents the type of association that would be expected when a random sample of individuals is taken from a population.

However, correlations obtained in this manner may well be irrelevant to what happens under extreme conditions. As emphasized in the previous chapters, extremes will result in high levels of mortality in populations. The genotype that has a relatively high fitness under these conditions will be extreme in its phenotype, and may be uncommon in a population. It is not clear if correlations based on average performance of all genotypes in a population can be applied to such genotypes. For instance, if an adaptive response to extremes occurs via a major gene, correlated responses may be different to those associated with a polygenic response.

The species border problem

One area where evolutionary constraints are obvious is in limits to the distributions of species. Many species have restricted distributions with clear-cut borders. This leads into the question of why such borders exist. Why don't organisms continue to evolve and therefore keep extending their borders forever? What constraints act to keep species within their current boundaries?

When borders are determined by physical barriers that prevent a species from extending its range, answers to such questions are obvious. The distribution of terrestrial species will be restricted by lakes and seas, while the distribution of species from low-lying marshlands will often be restricted by major increases in elevation.

However, borders commonly occur where abrupt changes in physical features of the environment are not evident. In these cases, ecologists search for more subtle changes at borders. Such changes may involve physical features of the environment such as climatic variables, biotic features such as competitive interactions and predation, or interactions between abiotic and biotic factors (Hoffmann & Blows, 1994).

Two types of evidence suggest that factors associated with climatic extremes are often important in limiting distributions. First, many species borders directly correlate with extremes such as the highest or lowest daily temperature of the hottest/coldest month, or the mean monthly rainfall of the wettest or driest month. Such correlations do not necessarily indicate that rainfall or temperature directly determine borders, but suggest that factors associated with extremes are important. For example, the distribution of kangaroo species in Australia is closely correlated with hottest and coldest temperatures as well as with annual precipitation, and this correlation is likely to reflect the effects of these climatic variables on food availability rather than direct selection via climatic extremes (Caughley *et al.*, 1987).

Second, there are many examples, particularly among plants and animals that are ectothermic, where historical changes in species distributions track climatic change. Ford (1982) described several animals that have altered their range in response to temperature changes. For example, the white admiral butterfly was widespread in southern England, but became restricted to the extreme south around the middle of the last century. This has been followed more recently by a northward spread, as warm summers reduced the duration of the larval and pupal stages of admiral butterflies which are susceptible to predation. In this example, climatic effects appear to be exerted through a biotic factor, so the effects of climate change are indirect.

Traits responsible for species borders have been identified in several cases. Gilbert (1980) studied populations of the aphid, *Masonaphis maxima*, which infects thimbleberry plants. This host plant has a wide distribution in North America, ranging from Alaska to Mexico. In contrast, the aphid's distribution is much narrower, extending from Vancouver Island in southern Canada to California. The distribution of *M. maxima* is therefore not limited by the distribution of its food plant. Instead, Gilbert found that its range was determined by the aphid's development time. *M. maxima* has to pass through three generations before sexual females arise, and only these females can produce eggs that are able to survive winter conditions. Populations must therefore undergo at least three generations a year to persist. Aphids cannot survive once their host plant sets fruit. In northern areas, the suitable plant tissue is not available long enough to allow for the requisite three generations.

As another example, Cooke (1977) has identified traits limiting the distribution of rabbits in Australia. Rabbits breed in the southern part of Australia but not in the northern part. The range limit is close to the Tropic of Capricorn, although isolated colonies occur further north. This limit corresponds to a change from rainfall occuring mainly in winter to a summer rainfall pattern. Rainfall is essential for new pasture growth needed by rabbits for breeding. Because rabbits have evolved in an area of autumn–winter rainfall in France and Spain, the fertility of males and mating behaviour of females is influenced by daylength. By becoming fertile prior to winter rainfall, rabbits improve their breeding success in these areas. However, this behaviour is not suitable for northern Australia with summer rainfall patterns.

From an evolutionary perspective, such ecological information can only provide a partial answer to the question of why particular boundaries occur. These studies identify the traits involved in limiting the range of a species, but do not indicate why such traits are limitations. For this purpose we need to investigate evolutionary constraints restricting species ranges. Why don't thimbleberry aphids evolve to produce sexual progeny at an earlier stage? Why don't these aphids evolve to develop more rapidly, or hatch earlier after winter, or use alternate host plants? Why don't rabbits in northern Australia evolve a different reproductive response to daylength? To answer such questions, we first need to look at the genetic level for the causes of species borders.

The most obvious of several hypotheses to account for borders at the genetic level (Table 4.2) is to assume that genes allowing an organism to adapt beyond borders do not normally occur. If development time needs to be more rapid for a border to expand, it may simply be that genes that further decrease development time do not arise in a population. In terms of the concepts we introduced in Chapter 2, borders represent situations where there is no available genetic variance for traits restricting the distribution of a species.

There are a number of possible reasons for this. First, there may be some sort of constraint at the physiological, biochemical or structural level that prevents the appearance of genotypes adapted to conditions beyond borders. An organism may simply be unable to decrease its development time or increase its resistance to a stress any further than it already has. Enzymes that allow an organism to function under a greater range of temperatures might simply not

Table 4.2. *Evolutionary hypotheses to account for limits to species ranges and some of their predictions*

Hypotheses	Predictions[1]			
	a. Low GV	b. Low h^2	c. Geogr. variation	d. Trade-offs
1. Physiological constraints/low heritability as a consequence of directional selection	✗	✓(field+ lab)	✓	✗
2. Low overall levels of genetic variation in marginal populations because of small population size	✓	✓	✗	✗
3. Traits show low heritability because of environmental variability in marginal areas	✗	✓(field only)	✗	✗
4. Favourable genotypes in marginal populations swamped by gene flow from central populations	✗	✗	✓	✗
5. Changes in independent characters required for range expansion	✗	✗	✓	✗
6. Adaptation to infrequent stressful conditions limited by genetic tradeoffs between fitness in favourable and stressful environments	✗	✗	✗	✓(environments)
7. Adaptation prevented by genetic tradeoffs between fitness-related traits in marginal populations	✗	✗	✗	✓(traits)
8. Accumulation of mutations deleterious under stressful conditions prevents adaptation	✗	✗	✗	✗

Note:
[1] Predictions relate to low overall levels of genetic variation (a), low heritabilities for traits under directional selection at borders (b), geographic variation among central and marginal populations (c), and tradeoffs between environments or traits (d).

exist. The structure of an animal's teeth or jaws may limit its ability to utilize food items outside its normal range. The nature of a plant's root system may restrict its ability to extract nutrients or water from infertile or dry soil. This explanation is commonly invoked to account for the existence of species borders.

Second, levels of genetic variation in populations at species borders may be severely reduced because these populations are often very small. Species often have a low density at borders as conditions become unfavourable, leading to populations that are persistently small or highly variable in size. This can have the effect of decreasing genetic variance in a population through the process of genetic drift. As a result of this process, genes that might be favoured under marginal conditions are lost by chance from a population. We can illustrate the effects of genetic drift by a simple example. Consider a bird population where a new gene has arisen that allows nestlings to survive cold conditions. This gene will initially be at a low frequency but is expected to increase due to selection. If the population becomes subjected to a sudden drop in size because of a reduction in food availability, only one or two pairs of individuals successfully breed. If by chance these individuals do not carry the gene enabling nestling survival under cold conditions, this gene will be lost from the population. Such a loss becomes more and more likely as populations persist for prolonged periods at a small size. In addition, a small size can result in inbreeding because related individuals will tend to mate with each other.

A third explanation for the absence of genetic variance at borders is that genetic variation is present, but not expressed because of environmental variability. Recall from Chapter 2 that heritability is determined by both the genetic variance and the environmental variance. If the environment is variable, this can both increase the environmental variance and decrease the heritability. In addition, environmental variation can directly influence the expression of differences between genotypes as we discussed in Chapter 2. Both processes may influence the ability of a population to respond to selection.

Borders may also exist because of reasons unrelated to levels of genetic variance. Alleles allowing range expansion may arise, but may be continually diluted by an influx of alleles from populations away from the border. If populations at borders are relatively small compared to those at the centre of species distributions, it is likely that one-way migration from central populations to marginal populations will predominate because of the large number of individuals in central areas. In this case, individuals carrying alleles that might be adapted to marginal conditions will often mate with those carrying different alleles that are favoured in central areas but not in marginal areas. Adaptation becomes difficult because favoured genes are continually diluted in marginal areas.

These different possibilities are further complicated by the likelihood that range expansion requires simultaneous changes in a number of traits, representing the sixth hypothesis in Table 4.2. Several environmental factors are usually correlated with species ranges, and a complex of adaptations may be required for range expansion to counter different factors. Plants that colonize mine tailings contaminated by heavy metals often have to deal with low nutrient levels as well as heavy metals. Insects in cold environments need to overcome the short length of the growing season as well as the low temperatures themselves. Animals in arid areas have to counter dry conditions and heat stress as well as a limited availability of food.

Evolutionary explanations of species borders are also complicated by tradeoffs (hypothesis 7). If strong tradeoffs occur between fitness in favourable conditions and adverse environmental conditions that sometimes occur at borders, these could constrain range expansion. Genes enabling species to expand their range may not increase in frequency in marginal populations because they are only favoured for brief periods during stressful conditions.

Finally, species borders may exist because of the accumulation of deleterious mutations in populations (hypothesis 8). The deleterious effects of many mutations may only be expressed in some environments and not in others. Those mutations with deleterious effects in the environment commonly experienced by an organism will be removed by selection. However, following an argument presented by Kawecki (1994), if environmental conditions determining borders occur only rarely, mutations that are deleterious under these conditions will accumulate in a population. Once the stressful conditions occur, the deleterious effects of these mutations may be sufficient to prevent a population adapting, thereby limiting range expansion. This hypothesis becomes particularly attractive if the deleterious effects of mutations are normally enhanced under stressful conditions, as suggested by *Drosophila* data collected by Kondrashov & Houle (1994). If the effects of mutations are environment-specific, there should be no fitness correlation between environments.

Species borders: genetic data

In theory, we should be able to test genetic hypotheses about species borders because a number of predictions can be made about specific hypotheses (Table 4.2). These predictions can be tested by genetic studies on ecological traits likely to be associated with the borders. If such traits do not exhibit genetic variance in marginal populations, this would suggest that species borders represent a genetic/physiological limit. If there is genetic variance, gene flow or interactions between traits may be important.

Unfortunately, little genetic information is available on species borders, making it difficult to evaluate hypotheses, even though this information could be obtained. We only really have information in three areas, namely geographical patterns, transplant experiments with plants, and comparisons of overall levels of genetic variation. We will discuss each of these in turn.

1. *Variation in traits between central and border populations*

In the absence of much ecological information on traits determining borders, we can obtain some indication of traits likely to be under selection at borders from geographical comparisons. If populations from different locations are reared under identical conditions, differences between them are likely to be genetic. Traits that change from central to marginal populations at borders could be under selection at borders, and therefore limit further changes to species distributions.

We can illustrate this approach with an example. The Queensland fruit fly, *Bactrocera tryoni*, has a wide distribution in eastern Australia, ranging from northern Queensland to south-eastern Victoria. This species evolved in rainforests where it utilizes soft fruit for breeding, and became a pest as it spread into orchards. *B. tryoni* was initially confined to Queensland, but has extended its range south since the mid 1800s. This expansion has involved evolutionary changes. Bateman (1967) examined the resistance of flies to high and low temperature extremes and compared strains from central populations and those at borders. Strains from a border population in Victoria took longer to die at both temperature extremes than those from other populations, while strains from northern Queensland were the most sensitive (Table 4.3). This is consistent with the fact that flies experience more extreme temperatures in southern locations. Resistance to temperature extremes (in particular low temperatures) could therefore be limiting further range expansion by *B. tryoni*.

As another example, we consider geographical variation in a cactus, *Opuntia fragilis*. Loik & Nobel (1993) compared the freezing resistance of populations of this cactus from North America throughout the range of the species. Plants were grown in controlled environments in an attempt to minimise the importance of non-genetic factors. Plants from populations at northern margins exposed to the coldest temperatures were more resistant than those from southern areas. Increased freezing resistance has therefore probably been selected in northern locations, and this trait may be associated with the northern border of *O. fragilis*. In both this study and in *Bactrocera tryoni*, we would need to assess genetic variance for temperature resistance to obtain further information about factors determining borders of these species.

2. *Patterns of genetic variation (mostly* Drosophila*)*

As we mentioned earlier, some information about factors determining species borders can be obtained from surveys of genetic variation in populations (Table 4.2). If small population sizes and inbreeding effects contribute to borders by reducing overall levels of genetic variation and causing the expression of deleterious alleles, this should be detectable in comparisons

Table 4.3. *Resistance of four populations of* Bactrocera tryoni *from eastern Australia to temperature extremes*

Numbers represent estimated time taken for 50% of the flies to die.

Population	Latitude (°S)	0°C	37°C
Cairns	17	36.9	25.6
Brisbane	27	37.4	30.5
Sydney	34	40.1	32.4
East Gippsland	38	40.4	40.3

Source: After Bateman (1967).

of genetic variation between central and marginal populations (hypothesis 2 in Table 4.2).

There is some evidence for decreased genetic variation at margins in comparisons of the frequency of 'lethal' and 'semilethal' genes in *Drosophila*. As implied by their description, these genes cause drastic reductions in fitness when they are in the homozygous form. The presence of lethal genes can be detected by crosses using special stocks available for some *Drosophila* species. Comparisons have been made between the frequencies of these genes in several in central and marginal populations of *Drosophila* (Parsons, 1983a). For instance, the frequency of lethals is lower in marginal populations of *D. melanogaster* undergoing drastic seasonal changes in population size compared with populations from tropical areas where population size is more stable. A relatively low proportion of lethal and semilethal genes also tends to occur in ecologically marginal populations of *D. pseudoobscura*.

Comparisons of *Drosophila* populations from central and marginal areas show consistent differences in another type of genetic variation, the incidence of polymorphism in chromosome inversions. Genes that are located within an inversion tend to be inherited together as a block. This is because there is not much recombination between an inverted sequence of DNA and one that is not inverted. In populations that are polymorphic for an inversion, both the inverted and non-inverted sequences are present. We can therefore determine the frequency of a particular inversion in the same way as we determine the frequency of genes in a population. In almost all *Drosophila* species that have been looked at so far, inversions show less polymorphism in populations at species borders than in central populations (Brussard, 1984).

In contrast to these findings, there is little evidence that overall levels of genetic variation are limiting in marginal populations. Genetic comparisons of marginal and central populations have been carried out by looking at variation in proteins detected by electrophoresis. Such studies provide little or no evidence for low levels of genetic variation in marginal populations, especially in *Drosophila* species (Brussard, 1984). Comparisons of central and marginal populations of plants also indicate no consistent trends, both for variation detected at electrophoretic loci and for variation in morphological traits that are more likely to be under direct selection (Wilson *et al.*, 1991). Safriel, Volis & Kark (1994) provided evidence that populations at margins may even be *more* variable than those at the core of species distributions; in wild barley, marginal populations had a higher level of variability than central populations in 14 out of 18 traits, while in the chukar partridge, allozyme heterozygosity was relatively higher in marginal populations. Small population sizes therefore do not seem to be a common explanation for species margins, although we should emphasize that many of these comparisons have not considered whether marginal populations exist in adverse conditions (i.e, are ecologically marginal) instead of simply being at geographic margins.

Inversion polymorphisms and lethal genes are either rare or difficult to characterize in species other than *Drosophila*. Nevertheless, an understanding of why different types of genetic variation show different patterns in *Drosophila* populations might provide general insights into genetic explanations for species borders. One hypothesis advanced by Brussard (1984) and others is that selection does not favour genes being locked up in inversions at species margins. This is because marginal conditions are more likely to be unpredictable, resulting in different combinations of genes being favoured at different times. The genetic flexibility provided by an absence of inversions may therefore be favoured.

However, it is difficult to reconcile this hypothesis with the idea that marginal populations at borders are under directional selection because of limiting environmental conditions. As already mentioned, directional selection will probably decrease genetic variance for traits under selection. The role of constraints will therefore remain unclear until genetic analyses of traits likely to influence species distributions are carried out. Information is required on levels of genetic variation for such traits in marginal populations, and this variation should preferably be measured under field conditions. We also need to examine the way these traits interact with other traits, and the role of gene flow in restricting the appearance of favourable genes.

One attempt to obtain information on genetic variance in ecologically-important traits was made by Blows & Hoffmann (1993) who investigated desiccation resistance in *Drosophila serrata*. This species is distributed along the east coast of Australia as well as in New Guinea, and has a southern border around Sydney. It was thought that desiccation resistance might be associated with the *D. serrata* border because the distribution of *Drosophila* species in this region seems to coincide with their relative resistance to desiccation stress. When central and marginal populations of *D. serrata* were selected for increased desiccation resistance, populations started with the same mean level of resistance, but central populations responded to a much greater extent over the first 10 generations (Figure 4.7). This experiment therefore provides little support for an increase in genetic flexibility in marginal populations and is inconsistent with the hypothesis that chromosome inversions decrease genetic flexibility. However, it is not clear if a lack of genetic variance for desiccation resistance is responsible for the border. If this trait was under directional selection, levels of resistance should be higher in marginal populations compared to central ones, which is not the case (Figure 4.7).

3. Tradeoffs in plants

Some plant studies suggest that tradeoffs may be important at borders (hypothesis 7 in Table 4.2). An advantage of using plants is that they can be transplanted between marginal and central areas of distributions to see if populations are adapted to local conditions.

As an example of this approach, consider the study by Schmidt & Levin (1985) on the fitness of the winter annual, *Phlox drummondii*. Fitness was measured as the 'intrinsic rate of increase' of a population, which is a measure reflecting the survival of plants as well as the number of seeds they produce. Rates of increase at a particular site tended to be highest for plants that originated there as mentioned in Chapter 3, suggesting that the populations were adapted to local conditions. Overall rates of increase were much lower in the unfavourable marginal areas than in central areas, in agreement with expectations. However, plants from marginal sites also had low rates of increase in favourable habitats when compared to plants originating from those areas (Figure 4.8).

This difference is probably related to growth rate. In unfavourable conditions, plants that have low growth rates tend to be favoured because they can conserve nutrients and other resources. In contrast, high rates of growth under favourable condition can enhance competitive ability in sites where plant densities are likely to be high. These findings and data on other organisms (Hoffmann & Blows, 1994) suggest that growth rate could generally form the basis of a tradeoff between environments. Because low growth rates will often be favoured under stressful conditions, tradeoffs between environments could contribute to species borders.

Species borders: indirect approaches

As mentioned above, it is often thought that genes enabling a species to expand its borders may not arise in populations because of physiological or biochemical constraints.

When these types of constraints occur, organisms are unable to obtain

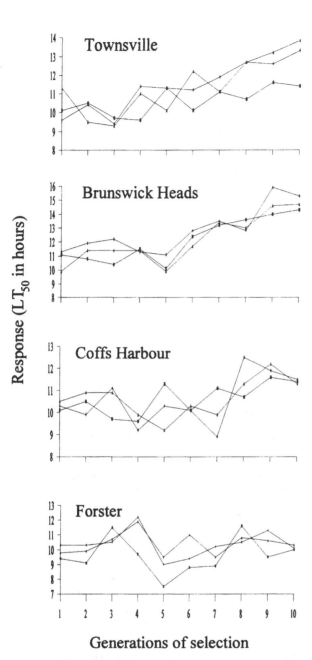

Figure 4.7. Response of four populations of *Drosophila serrata* from eastern Australia to selection for increased resistance to desiccation. Selection was carried out on three independent replicate lines from each population for 10 generations. The response is given as the time taken for 50% of the flies to die (LT_{50}). Populations are given in order of their proximity to the southern Australian limit of this species at Forster, with Townsville being at the centre of its distribution. (From Blows & Hoffmann, 1993).

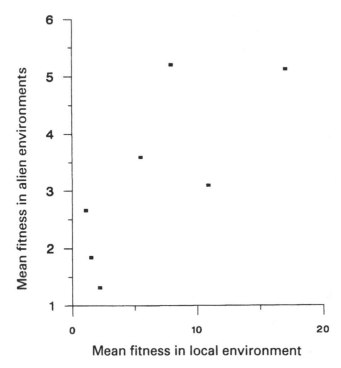

Figure 4.8. Mean fitness of populations of an annual plant, *Phlox drummondii*, in their own environment plotted against their mean fitness in six other environments. Fitness is measured as the rate of increase in size of populations. Genotypes from less favourable environments (lower rates of increase) also seem to have a lower fitness in alien environments. (From data in Schmidt & Levin (1985)).

sufficient resources for survival or reproduction. This will often result in organisms having insufficient energy. One reason for this is that organisms normally suffer large metabolic costs when they are under stress. Examples include the increases in metabolic rate that occur when mammals are exposed to pollutants or temperature extremes, and in particular when they experience combinations of these stresses. These demands at species borders are likely to restrict the energy available for other activities, which can be very expensive in terms of energy, requiring almost all of the energy intake of an animal (Peterson, Nagy & Diamond, 1990). Other resources, such as the availability of nitrogen which is essential for the production of proteins, can also be generally limiting (White, 1993).

We may be able to determine the importance of energetic constraints or other constraints in limiting species distributions without collecting genetic data. If the same constraints limit the distribution of different species, then these constraints should be associated with the borders of unrelated species.

An example of this approach is provided by Root's (1988) work on the northern borders of many different birds in North America. Borders of different species often followed isotherms for the mean minimum January temperature, suggesting that a factor related to temperature was involved. Root examined the metabolic rate of several of the species when they were resting. For each species, the resting metabolic rate at the temperature corresponding to the mean minimum January isotherm of its border was 2.5 times higher than the lowest metabolic rate (basal rate) that was achieved. Thus, while the borders of species were associated with different isotherms, their distributions were associated with the same physiological factor. This relationship seemed to hold regardless of the degree to which birds were related, and for variations in body size and the types of habitat utilized.

The 2.5 generalization may be related to the amount of energy birds need to keep warm. The resting metabolic rates of birds at cold temperatures are higher than basal rates because individuals have to expend energy. Species borders may therefore coincide with the amount of energy birds need to expend beyond the basal rate in order to keep warm. However, we should emphasize that the universality of the 2.5 generalization has been questioned. Repasky (1991) argued that this rule may not apply to many birds. Instead, he suggested that food availability in addition to temperature stress is likely to be involved in the northern borders of many species. These factors will have similar effects because both temperature extremes and food availability will influence the energy balance of birds.

In summary, we still know very little about factors determining borders at the evolutionary level. Constraints and tradeoffs may be involved, but there is insufficient genetic information on traits determining borders and their interaction with other traits. Non-genetic data can also provide information about evolutionary constraints, but we do not know if borders of different species will tend to be determined by the same constraints.

Humans: stress and life at the margins

Our own species can occupy hostile environments to some extent because of our ability to manipulate them, although this ability has developed relatively recently. Ultimately *Homo sapiens* has been exposed to the same selective pressures as other species in adjusting to natural environments. It is likely that early forms of our species originated in dry tropical savannahs. These early forms were probably adapted to rather high day time temperatures, cool (but above freezing) night time temperatures, moderately low humidities, high solar radiation, and a relatively high oxygen pressure found at low altitudes.

Can we find evidence of adaptation to environmental extremes, and to what extent is our current distribution limited by extremes? A difficulty in studies on our own species is that we cannot carry out the types of experiments available in animals. Studies of populations in identical environments cannot be carried out, in comparison with experimental animals where populations can be replicated and environments controlled. This means that interpretations about the genetic basis of variation in human populations is often imprecise. However, human data sets can be enormous and cover a wide range of populations and habitats. We can use this data set to investigate associations with climatic variables.

In humans, temperature is regulated to maintain a core body temperature of around 37°C, although factors such as age, sex, and time of day lead to some variation around this value. Extremes of cold and heat can cause death directly in humans, although indirect effects can also be important. For example, a study of the association between temperature and mortality from coronary heart disease and stroke in American cities (Rogot & Padgett, 1976) showed minimal mortality between about 15°C and 27°C, with increases in mortality above and below this range. In addition, very hot days exerted a cumulative effect on mortality. High mortality rates have also been associated with short spells of relatively hot weather in London (Haines, 1991).

In order to tolerate hot environments and maintain body temperatures, humans have evolved a high density of sweat glands with a high capacity for heat loss. Humans seem to be superior to other primates in their capacity to sweat, and have a heat loss capacity in excess of their ability to produce heat through working (Hanna & Brown, 1983). Heat loss is most efficient in dry environments, and elevations in heat level alone are less stressful than a combination of heat plus high humidity. This is because evaporation, which is one way of dissipating heat, is restricted under humid conditions. In an environment that is both hot and dry, evaporation for removing heat is more effective than under humid conditions when sweating loses its efficiency.

Heat resistance can be modified by acclimation which can lead to substantial variation among individuals. Adults can be rapidly acclimated by periods of work in hot, humid environments (Hanna & Brown, 1983). Acclimation results in a decrease in heart rate and rectal temperature, and an increase in sweat rate. In addition, long-term 'developmental' acclimation occurs in children. When individuals are exposed to higher heat loads during childhood, a greater percentage of the sweat glands are likely to be activated. This means that the climate in which an individual develops may have some influence on adult heat resistance.

Turning to cold, our tropical origins mean that resistance to cold should be low in spite of our large body size. We respond to cold conditions by increasing our metabolic rate, and without clothing our response is much more pronounced than in native arctic mammals which have significant surface insulation, but our response is similar to tropical mammals (Harrison *et al.*, 1988). Following cold exposure, homeostatic responses such as shivering can increase metabolic rate 100% above basal metabolic rate. Activities such as running can increase heat production much more, but the length of time that high metabolic rates can be maintained in the face of cold stress is quite short. Furthermore, human populations have a limited acclimation response to cold as compared with substantial ability to acclimate to heat.

When humans are exposed to cold temperatures, they undergo a series of reversible changes, starting with a loss of mental alertness and eventually leading to an unconscious or semi-conscious state at core temperatures of 3 or 4°C below normal. Conditions producing unconsciousness in adults depend on the amount of fat. Individuals who are nude, inactive and thin can become unconscious when they are exposed to still air temperatures of 5°C for more than 2 hours, while those with heavy fat layers can remain fully functional under these conditions. As in the case of heat stress, children are more susceptible to cold stress than adults, which partly reflects their smaller size.

Experiments on heat resistance have been carried out on various groups including Europeans, Australian Aboriginals, North Africans and South African Bantu. Major differences in resistance occur when groups are first tested under hot working conditions, but these tend to become minimal after heat acclimation. This suggests that there may not be much genetic difference among these populations for physiological resistance to heat stress (Harrison *et al.*, 1988). Any remaining differences tend to be associated with variation in body size, physical fitness, and levels of subcutaneous fat.

To examine adaptive differences in body size and shape, attempts have been made to correlate these measures to environmental conditions in different ethnic groups. Genetic factors are suggested when groups differ when they are measured under the same environmental conditions. Two examples of such relationships are given in Figure 4.9, which give the relationship between mean annual temperature and both weight and basal metabolic rate in several groups. For weight, there is a decrease as mean environmental temperature increases. This trend is expected because small body size can increase heat resistance by providing a larger surface area for cooling, whereas a large body size can reduce heat loss. The weight and metabolic rate differences between ethnic groups may be adaptive. For instance, Amerinds may have originated from groups exposed to cold climates, accounting for their relatively high weights. Turning to metabolic rate, there is a decrease in metabolic rate from the pole to the equator as heat stress increases. Substantial differences among populations occur. Mongoloid peoples have the lowest metabolic rates and Amerinds the highest. The decrease in metabolic rate towards the equator could be adaptive because heat resistance is associated with a low metabolic rate. This trend mirrors acclimation responses to heat stress which result in a lowered metabolic rate.

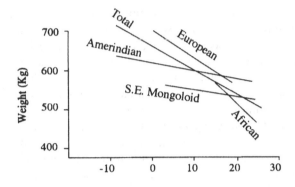

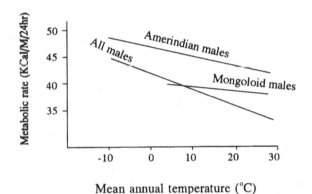

Mean annual temperature (°C)

Figure 4.9. Relationship between mean annual temperature and (*a*) mean body weight and (*b*) basal metabolic rate in human ethnic groups. Lines represent regression lines indicating how temperature affects weight and metabolic rate. (From Harrison *et al.* (1988) and MacFarlane (1976)).

While these studies suggest that differences among ethnic groups can be related to the types of climates they experience, there are problems in attempting to make generalizations solely on the basis of broad associations with temperature. Houghton (1990) considered the case of Polynesian people. Most Polynesian islands lie in the tropics, so small individuals are expected on the basis of the temperature–weight association in Figure 4.9. However, Polynesians from small islands tend to be tall and muscular, which would seem to suggest that they live in a cold climate. Houghton pointed out that windy and wet conditions are frequently encountered in oceanic environments of Polynesia. Under these conditions, human bodies can chill rapidly and be exposed to a cold stress. Moreover, technology available to these people provided little protection from this environment. The importance of cold selection is highlighted by the fact that people living inland in larger islands of the Pacific Ocean often do not possess the large muscular forms of Polynesians from more exposed maritime environments.

Apart from adaptation to temperature extremes, research in humans has also focussed on adaptation to high altitudes. Permanent settlements can occur at altitudes up to 5000m, although high-altitude populations suffer fitness reductions as indicated by reduced fertility and growth, increased congenital malformations, and an increased level of mortality and sickness (Moore & Regensteiner, 1983). The physiological effects of living at this elevation are related to the oxygen binding characteristics of haemoglobin. Blood leaving the lungs is nearly saturated with oxygen at 2000m, but by 4000m blood oxygenation is substantially lowered. The immediate response to a lack of oxygen (hypoxia) is an increase in breathing and heart rates. This can be partially overcome as people become acclimated, particularly if acclimation occurs from birth (Frisancho *et al.*, 1995).

There is some evidence that high-altitude populations are adapted to hypoxic conditions. People from the Andes and Himalayas have higher maximum oxygen capacities than individuals from low altitudes who have moved into these environments despite the fact that the latter have undergone acclimation. Hochachka *et al.* (1991) explored the physiological basis of this difference in people from the Andes. They found that Andean people accumulated less lactate under hypoxic conditions than individuals from low altitudes. Lactate normally accumulates when metabolism occurs through inefficient pathways that do not require oxygen. In contrast, lactate does not accumulate when metabolism proceeds via efficient pathways that utilize oxygen and produce more energy for activity. This difference may therefore help to explain why high-altitude peoples can maintain high levels of activity despite hypoxic conditions. Hochachka *et al.* (1991) also tested whether the difference between groups disappeared once Andean people moved to lower latitudes. They found that Andean people continued to metabolize efficiently under hypoxia even after six weeks at a low altitude. Differences between groups may therefore involve long-term acclimation, perhaps due to developmental changes, or else genetic differences between populations.

This brief discussion suggests ways in which human populations may have directly adapted to environmental conditions. Humans seem well adapted to hot conditions but not to cold conditions. Cold temperatures and a lack of oxygen will often limit our distribution. There is some evidence for adaptive changes to climatic conditions. However, except in extreme situations, it remains difficult to conclusively demonstrate that differences among human populations are genetic, because of cultural factors influencing nutrition and technologies that may have a large impact on resistance to climatic extremes.

Summary

Tradeoffs may occur as organisms adapt to new environmental conditions. Biochemical and physiological considerations suggest that trade-offs between environments should be common. Several studies have demonstrated the existence of tradeoffs by focussing on specific traits or on specific genes. Increased fitness in stressful environments may often be associated with decreased fitness in favourable environments.

The detection of tradeoffs is complicated by acclimation and coadaptation. Organisms may adapt to new conditions by changing their ability to become acclimated, but this can result in a tradeoff with other conditions. Coadaptation occurs when genetic interactions influence fitness. Fitness costs due to a major gene may be decreased by coadaptation with modifier genes.

Evolutionary explanations for species boundaries differ from ecological explanations in that the former consider why species cannot adapt to conditions outside borders whereas the latter consider traits and variables associated with borders. A number of hypotheses have been proposed but there is not much relevant empirical data. Physiological and biochemical constraints may be important in restricting range expansions into harsh environments. Stress response traits of direct ecological significance need to be considered in determining the possibility of range expansions, rather than variation at the direct genetic level such as allozymes, inversions and lethal or deleterious genes.

Early hominid forms evolved in dry tropical savannahs, and humans have a greater ability to acclimate to hot and dry conditions than to cold and humid conditions. Limits to the distribution of human populations involve heat (especially under humid conditions), cold, and hypoxia at high altitudes. Increased metabolic rate tends to be an immediate response to such stresses especially in newborns, although acclimation can rapidly ameliorate some responses. However, the human species cannot survive and reproduce beyond certain limits even though some extensions may occur via cultural changes.

CHAPTER 5

Evolutionary outcomes: comparative and optimality approaches

We have so far considered the effects of natural selection on evolutionary changes over a relatively small timespan. In some cases, selection
was monitored as evolution occurred. In others, the effects of selection were
inferred indirectly by demonstrating divergence among populations.

In this chapter, we consider inferences about outcomes of selection over a
longer timespan. Two approaches have been used in this context. The first constitutes the 'comparative approach', involving comparisons among species
from different environments. This approach has been used extensively to look
at divergence in traits ranging from the morphological level to the biochemical level. Recent developments in the comparative approach allow the degree
of relatedness among species to be considered when determining whether
adaptive changes have occurred. The second approach tests for the presence
of characteristics which species are expected to have evolved when exposed to
particular conditions for a long period. This constitutes the 'optimality
approach', and is being used to examine how organisms might be expected to
evolve when they encounter extreme as well as favourable conditions.

We discuss limitations of the comparative and optimality approaches in
making deductions about adaptive changes in populations. Both approaches
do not usually address the process of evolution because they are largely concerned with its outcome. This means that they can only examine the types of
evolutionary responses that organisms have made when faced with environmental extremes, and not the importance of extremes in driving evolution.
Nevertheless, the comparative and optimality approaches are important tools
in identifying the types of traits that may be involved in responses to extreme
conditions.

The comparative approach: divergence among species

There are many reasons why related species occur in different environments. In some cases, one species may exclude another because of biotic
factors such as competition. For instance, the southern distribution of the
arctic fox (*Alopex lagopus*) is determined by its ability to compete with the red

fox, *Vulpes vulpes* (Hersteinsson & Macdonald, 1992). In many other cases the distribution of species can be related to the direct and indirect effects of climate (Hoffmann & Blows, 1994). The absence of one species from an environment will often reflect the fact that the environment is too extreme to enable it to survive and/or reproduce as discussed in the previous chapter.

When a species is excluded from an area because of the extreme nature of the environment, comparisons can be made between related species to identify traits that might be involved in coping with particular conditions. If species differ for a particular trait, this difference may enable species to exist in their respective environments. Species differences could, in theory, also arise because of the random process of genetic drift (see Chapter 1), or because species have become adapted to environmental conditions in the past which bear little resemblance to conditions they now experience. These factors will become more important as species evolve independently for a longer time because the effects of random drift and chances of encountering different environments will have increased. It is therefore advisable to use closely-related species occupying different environments in such comparisons.

Many comparisons between species focus on a pair of related species, and we illustrate this approach with two examples. The first considers differences in the enzyme malate dehydrogenase from two species of cattails (*Typha latifolia* and *T. domingensis*) studied by Liu, Sharitz & Smith (1978). These plants occur in aquatic and marshy areas in temperate and tropical areas. Malate dehydrogenase (Mdh) is important in the metabolism of organisms and involved in the oxidation of carbohydrates. Liu *et al.* (1978) considered populations of the two species growing in Par Pond, a lake in South Carolina partly heated by effluent from a nuclear reactor. Discharges in the lake produce a thermal gradient with a maximum temperature in the range 32–35°C. *T. latifolia* grows in warmer areas of the lake than *T. domingensis*, and the latter species does not occur when maximum water temperatures exceed 30°C, suggesting that this environment is extreme for *T. domingensis*. To examine differences in the thermal stability of malate dehydrogenase (Mdh), this enzyme was extracted from leaves and exposed to different temperatures to see when its activity was lost. Mdh from *T. latifolia* maintained its activity above 35°C, whereas Mdh from *T. domingensis* did not. This suggests that differences in the distribution of these species are consistent with the thermostabilities of one of their enzymes. Selection may therefore have been responsible for divergence between Mdh from these species, and Mdh differences may contribute to the ability of one of them to survive under more extreme conditions.

A second example that focusses on the physiological rather than the bio-chemical level is a study by Asami (1993) on desiccation resistance in two species of land snails, *Mesodon normalis* and *Triodopsis albolabris*. Adults were collected from the field as juveniles and raised in the laboratory. Juvenile offspring from these parents were then tested for resistance to desiccation and

(a) Survival rates

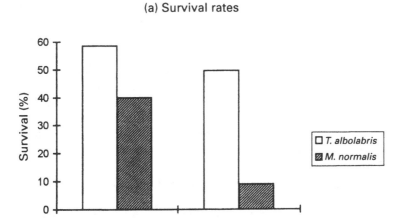

(b) Dehydration rates

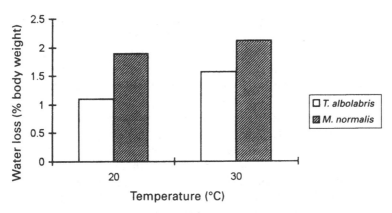

Figure 5.1. Differences in survival and dehydration of juveniles of two species of snails, *Mesodon normalis* and *Triodopsis albolabris*. (a) Survival rates after 22 days under desiccating conditions at two temperatures. (b) Rate of dehydration in live snails after desiccation for 22 days at two temperatures. (Modified from Asami, 1993).

rates of water loss. Because tested individuals had been raised in the same environment, differences between the two species were likely to be genetic rather than environmental in origin.

Asami (1993) found that *M. normalis* was less resistant to desiccation than *T. albolabris* when juveniles had been desiccated for a few days (Figure 5.1(a)).

This difference is probably not associated with the types of shell possessed by these species because they have similar shells as juveniles. Instead, the rate at which the species lost water via evaporation was probably responsible because this rate was much lower in *T. albolabris* than in *M. normalis* (Figure 5.1(b)). This physiological difference matches differences in the habitats that the species occupy. Although both species occur in the uplands of the Appalachian mountains in the U.S.A., *T. albolabris* has a much wider distribution and occurs at low altitudes down to sea level. In these low areas in summer, snails are exposed to high temperatures and desiccating conditions as litter on the ground becomes dry, and these conditions appear to be too extreme for *M. normalis*.

Although such studies suggest that adaptive responses to extremes have occurred, the comparative method based on species pairs is limited. Because each species represents an independently evolving unit, associations are based on only two data points, and could easily have arisen by chance, unless multiple pairs of species are tested. We can be more confident that differences between species pairs reflect adaptive changes by making specific *a priori* predictions about the differences we expect on the basis of the environments being compared. In the snail example, predictions about desiccation resistance and water loss could be made from the distribution of the two species. Differences in the thermal stability of Mdh in cattails were also consistent with environmental predictions, although it is difficult to conclusively show that biochemical variation is responsible for an organism's ability to survive and reproduce in an environment. In other words, variation in Mdh may only make a minor contribution to survival differences between the species even though this enzyme is central to metabolism.

To overcome the limitations of using only two species, a group of related species can be examined to provide more data points for testing environmental associations. If the same values of a trait occur in several species exposed to similar environments, this makes it more likely that variation in the trait is adaptive. We illustrate the approach with two traits that are likely to be important in adaptive responses to environmental extremes, namely the production of a particular class of proteins and metabolic rate.

A small number of proteins known as the 'heat shock proteins' (hsps) are rapidly synthesized by organisms or cultured cell lines when they are exposed to high temperatures. The same proteins can also be induced by a number of other stresses, which has led to speculation about their role in generally protecting organisms and cells from adverse conditions. It is believed that hsps protect other proteins from breaking down as a consequence of stresses within a cell, enabling cells to become resistant. Several lines of evidence support this idea. For instance, the time when the synthesis of heat shock proteins takes place coincides with the time when cells acquire increased resistance to high temperatures.

Table 5.1. *Habitats of nine lizard species, temperature ranges in which the species are active, and temperature ranges in which hsp68 is induced and continues to be synthesized*

Species	Habitat	Temperature range of species activity (°C)			Temperature range for induction (°C)
		Minimal	Optimal	Maximal	
Phrynocephalus interscapularis	Sand desert	15–17	20–34	37–42	39–50
Phrynocephalus raddei	Sand and clay desert	11–14	17–30	38–40	39–47
Phrynocephalus helioscopus	Clay desert	7–10	15–30	36–42	37–47
Crossobamon eversmanni	Sand desert	10–13	16–23	24–29	37–45
Teratoscincus scincus	Sand desert	8–10	16–22	26–30	36–43
Agama caucasica	Rocky slopes and canyons	8–10	20–26	28–38	37–47
Lacerta agilis	Mountains, steppe, forests and gardens	7–10	20–25	27–35	36–42
Lacerta vivipara	Deciduous and coniferous forests	12–15	18–25	27–32	35–42
Lacerta saxicola	Rocky mountains	7–10	19–24	28–32	36–42

Source: Simplified from Ulmasov *et al.*, 1992.

Because of the likely role that these proteins have in enabling organisms to resist high temperatures and other stresses, researchers have sought associations between the quantity of protein produced by a species and its thermal environment. Ulmasov *et al.* (1992) tested for such an association in nine lizard species from diverse habitats, by focussing on one group of heat shock proteins. Lizards were exposed to heat and the proteins produced in their livers were then examined. Species from hot environments synthesized one of the stress proteins of this group (hsp 68) at the high temperatures they normally experienced (Table 5.1). In contrast, species from cooler environments (i.e., species likely to find hot environments extreme) synthesized hsp 68 proteins at lower temperatures, but which still exceeded temperatures they experienced in nature. This suggests that hsp 68 performs an important protective function for species from hot environments when hsp 68 is induced under natural conditions. As might be expected, species from hot environments also produced more of this stress protein after induction by a heat shock than those from cooler environments.

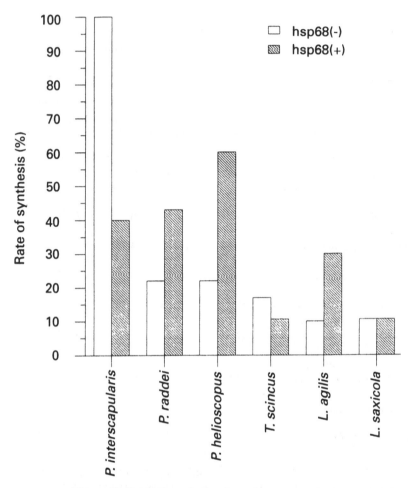

Figure 5.2. Rate of synthesis of two forms (+ and −) of a heat shock protein (hsp 68) in six species of lizards, arranged in order of the decrease in the average temperature of their habitat. The rate of synthesis of the hsp(−) form in *P. interscapularis* is arbritarily set at 100%, and other rates are expressed relative to this value. (From Ulmasov *et al.*, 1992).

Ulmasov *et al.* (1992) reasoned that species from hot environments might normally produce higher levels of hsps even under optimal conditions. They therefore examined the amount of hsp 68 produced by the lizards at a non-stressful temperature. As expected, there was an association between the mean environmental temperature a species experiences and the production of this protein (Figure 5.2). These species comparisons therefore strongly suggest that temperature conditions are closely tied to both the presence of hsp 68 proteins under non-stressful conditions and the amount of hsp 68 induced after a heat shock. However, as in the case of Mdh in cattails, the relative contribution that hsp 68 makes to species differences in the survival of extremes is not known.

Another example of this approach is provided by associations between environmental conditions and the metabolic rate of many animals. One way animals can counter adverse conditions is to lower their requirements for energy and other resources. The rate of energy metabolism of many species has been measured as the rate of oxygen consumption. Oxygen is required for most metabolic processes, and its consumption rate can therefore be used to monitor metabolism. Because metabolic rate depends on the activity of animals, oxygen consumption is normally measured in organisms which are not active, providing a measure of their 'resting' metabolic rate.

Many studies have compared the resting metabolic rates of animals restricted to benign environments with those from adverse conditions. In these comparisons, some compensation is usually made for differences in the body size of the animals, because large organisms have higher metabolic rates than small ones. Once oxygen consumption has been adjusted for size, species from adverse conditions often have lower resting metabolic rates than those usually found in more favourable conditions.

An example (Lovegrove, 1986) is provided by subterranean rodents which live in a variety of environments with different rainfall levels. In general, rodents found in dry areas have much lower resting metabolic rates than would be expected from their weight, whereas species restricted to wet areas have higher metabolic rates than expected (Figure 5.3). A lower metabolic rate means a reduction in the amount of energy used by the rodents while they are resting. This translates into a lower food requirement, which can be adaptive in dry environments where food is scarce.

Associations between metabolic rate and environmental adversity have been found in a variety of invertebrates including ants, beetles, corals and limpets (Hoffmann & Parsons, 1991). A low resting metabolic rate helps to counter many forms of environmental extremes including desiccation and high temperatures as well as food availability, although it may not be associated with the detoxification of specific toxins. In endotherms, a low rate will not always be selected under adverse climatic conditions, because a high metabolic rate enables endotherms to withstand cold conditions and thereby occupy a wider range of environments (eg. Mugaas, Seidensticker & Mahkle-Johnson, 1993).

When comparisons involve several species rather than a species pair, we can be more confident that traits have evolved in response to environmental conditions. However, data in the above examples were not corrected for the degree to which different species were related. This can be a problem when related species tend to occupy similar habitats. For instance, consider the situation where three closely-related species of lizards occupy one type of environment, while another type is occupied by three different lizard species that are closely-

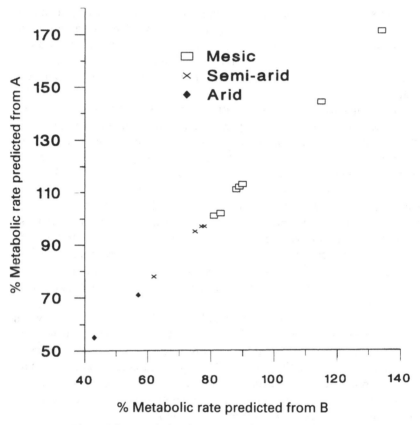

Figure 5.3. Association between resting metabolic rate and three habitat types in 13 species of subterranean rodents. The resting metabolic rate is presented as a percentage of that predicted from the association between metabolic rate and body size in rodents (A: plotted along the *y* axis) and subterranean rodents (B: plotted along the *x* axis). Data points indicate a resting metabolic rate for arid species that is lower than expected on the basis of either predictions. (Plotted from data in Lovegrove, 1986).

related to each other but distantly related to the first group. Because the related species diverged more recently in their evolutionary history, they are likely to be more similar to each other than to species from the other group. This means that lizards within a group are more likely to share characteristics unrelated to the types of environments from where they came. Instead of determining the association between environmental variation and a trait based on six independent data points, there may only be two independent data points representing two evolutionary pathways. Species comparisons should therefore ideally be adjusted for the degree of relatedness if adaptive changes are to be identified.

This problem is likely to apply to the examples we considered above. For instance, in the Ulmasov *et al.* (1992) study, the three species found in warm environments were from the same genus, whereas three species from cooler environments were from a different genus. It is possible that genetic divergence occurred between ancestral species that gave rise to these genera. This severely reduces the number of independent data points testing for an association, although both the associations with heat shock proteins and metabolic rate are consistent with physiological/ biochemical predictions.

Considering phylogeny in species comparisons

When information about the phylogeny of a group of species is available, this can be used to reduce the likelihood that relatedness among species results in spurious correlations between a trait and environmental variation. Phylogenetic information can also been used to investigate the direction in which evolution has occurred and determine which characteristics represent ancestral forms.

Phylogenetic data can be related to variation in stress resistance among *Drosophila* species belonging to the *melanogaster* species group. The resistance of six species from this group to a −1°C cold shock and to desiccation is plotted in Figure 5.4. *D. melanogaster* and *D. simulans* are both widespread and occur in a range of climatic conditions. *D. melanogaster* is the most resistant for both stresses, while *D. simulans* has a relatively high level of cold resistance. The other four species are restricted to the tropics and are relatively sensitive to both stresses. *D. yakuba* and *D. erecta* are particularly sensitive to cold. These data can be compared to phylogenetic relationships among the species. A phylogenetic tree determined from variation in a particular sequence of DNA indicates that four of the species tested are closely related, while *D. yakuba* and *D. erecta* are more distantly related to this group (Figure 5.5). Stress resistance is uncorrelated with the degree to which the *Drosophila* species are related because *D. melanogaster* and *D. simulans* are more closely-related to other species than to each other. This suggests that the high resistance levels shown by *D. melanogaster* and *D. simulans* have evolved recently and independently as these species became widespread and overcame climatic extremes.

Several attempts have been made to develop more objective and quantitative techniques of correcting for the degree of relatedness among species (Harvey & Pagel, 1991). For instance, consider the phylogeny for four species in Figure 5.6. Species 1 and 2 have evolved from a common ancestor (indicated by A). The divergence of these species from A represents an evolutionary event that is independent of another event involving divergence of species 3 and species 4 from B. Finally, the divergence of ancestors A and B from C represents a third event. We therefore have three independent evolutionary events

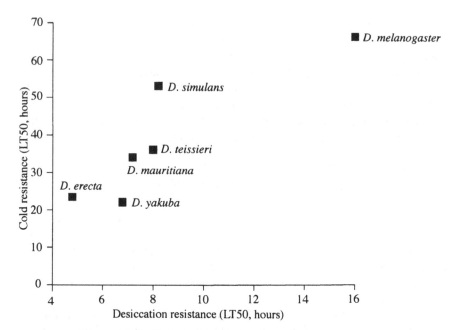

Figure 5.4. Resistance of six *Drosophila* species from the *melanogaster* species subgroup to desiccation and cold stress. Resistance is measured as the time taken for 50% of the flies to die (LT_{50}). (From Stanley *et al.*, 1980).

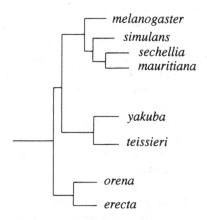

Figure 5.5. Phylogenetic relationship among eight *Drosophila* species of the *melanogaster* species subgroup as determined by DNA sequence homology in the gene coding for the alcohol dehydrogenase enzyme. The relationships among the closely-related species *D. simulans, D. sechellia* and *D. mauritiana* were not entirely resolved by this method. (Simplified from Jeffs, Holmes & Ashburner, 1994).

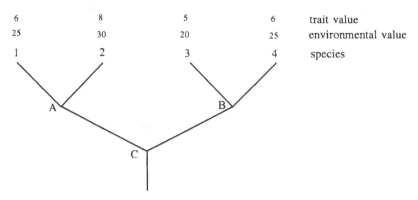

Figure 5.6. Hypothetical phylogeny between four species (1–4) and three ancestors (A–C) along with hypothetical values for a trait and an environmental variable. (Modified from Harvey & Pagel, 1991).

to account for the variation among the four species. In Figure 5.6, measurements are given for a trait and an environmental variable for each species. Is there an association between the environment and the trait? If we treat each species as an independent data point, there is a positive association between the trait and the environment. This might be taken as evidence that variation in the trait is adaptive. However, when we consider the three independent evolutionary events represented in this phylogeny, a different story emerges. While divergences between the two species pairs (1 vs 2, 3 vs 4) are associated with different trait values (6–8 vs 5–6), the environmental change is 5 units in both cases. We can estimate trait and environmental values for the ancestors (A, B) as the mean of the species they have produced. This also indicates an environmental change of 5 units (27.5–22.5), but a trait change of 1.5 units (7–5.5). An association between the trait and the environment is therefore no longer evident once the degree of relatedness among species is taken into account.

This technique was used by Stone & Willmer (1989) to examine the association between climate and body temperature in bees. These insects regulate their body temperatures to some extent by contracting flight muscles located in their thorax, allowing them to remain active under cold conditions. Species restricted to colder climates are expected to have higher warm-up rates than those from warmer climates. Differences in warm-up rates may be partly mediated by body mass. In small insects, the rate of heat loss is likely to be an important factor in determining warm-up rates, so larger insects with lower surface-to-volume ratios are expected to warm up more quickly than smaller ones. These hypotheses were tested by Stone & Willmer (1989) using data from 55 bee species. They measured the minimum air temperature at which bees were active, body mass, and the rate at which bees warm up. When associations between species were analysed using species means (i.e., without considering phylogeny) there was a significant negative correlation between warm-up rates and the minimum temperature at which bees were active. Species from colder climates therefore had higher warm-up rates as expected. In addition and contrary to expectations, there was no association between warm-up rate and body mass when species were compared. After controlling for phylogenetic relatedness by using the actual evolutionary changes themselves as data points in the analysis as described above, the negative correlation between warm-up rate and temperature for activity remained. However, this analysis also revealed the predicted association between warm-up rate and body mass. Phylogenetic considerations therefore revealed an association not evident from a direct comparison of species. Bees from colder climates appear to have evolved an ability to warm up quickly, and this has been aided by evolutionary changes in body mass.

As well as being used to investigate associations between traits and environmental conditions, phylogenies can indicate patterns of evolutionary changes. By determining the simplest way the distribution of a trait in a taxon could have evolved, information is obtained on the trait values and environmental conditions likely to have been ancestral.

This approach can be illustrated with an example. In the previous chapter, we discussed the case of wing polymorphism in water striders; in polymorphic species, individuals from areas where resources occur sporadically tend to have long functional wings, whereas those from permanent water bodies tend to have short, non-functional wings. Because some species of water striders are monomorphic for short or long wings (ie., only have one or the other wing type), a phylogenetic analysis can be used to examine the direction of wing evolution (Andersen, 1993). For instance, in the genus *Aquarius*, the pattern of wing morphs in diapausing and non-diapausing individuals from 10 species is illustrated in Figure 5.7. The simplest way to obtain this pattern is to assume that the ancestral state was monomorphic for short wings in both diapausing and non-diapausing individuals. This means that only two (for the non-diapausing state) or three (diapausing state) evolutionary changes are required to produce the wing patterns in the different species. Because short wings are ancestral, water striders must have originally inhabited permanent habitats where this type of wing was favoured, and adapted to changing habitats later in their evolutionary history.

We should emphasize that these types of analyses depend on a reliable phylogeny for the species under investigation. Unfortunately, phylogenetic relationships have only been determined for some groups of organisms, and these phylogenies may not always be accurate. Different taxonomic techniques and different types of taxonomic data applied to the same groups of organisms can yield inconsistent phylogenies. For instance, the phylogenetic position of giant pandas (*Ailuropoda melanoleuca*) is unclear and depends on the type of data that is used. Morphological measurements, immunological data and DNA

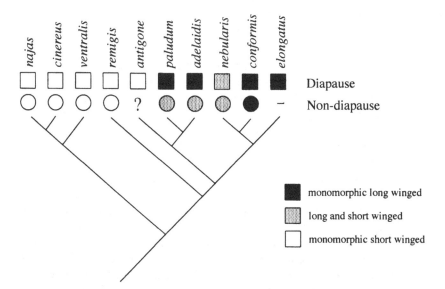

Figure 5.7. Phylogenetic distribution of wing morphs in ten species of *Aquarius* water striders. Black shading indicates long wings, unshaded indicates short wings, and patterned indicates polymorphic species. Data are presented for diapausing and non-diapausing adults. (Simplified from Andersen, 1993).

sequence data have all been used to determine relatedness between this species and other mammals. In some analyses, the giant panda has been placed in the bear family, while in others it has been placed in the raccoon family, or in a separate family with another panda species, the lesser panda (*Ailurus fulgens*). Pandas have even been placed in different groups when different taxonomic techniques are applied to similar data sets (eg Hashimoto *et al.*, 1993). Care should therefore be taken when making evolutionary inferences with a phylogeny based on one type of analysis of a single data set.

Tradeoffs and evolutionary constraints

In the previous chapter, we considered evidence which suggested that tradeoffs between environments often occur. Such tradeoffs can constrain long-term evolutionary changes, because adaptation to one environment can decrease fitness in another environment. If constraints exist, then the effects of tradeoffs should be evident when comparisons are made between species or higher taxonomic levels. Species adapted to one set of environmental conditions should be poorly adapted to another set of conditions.

One way of examining the effects of tradeoffs at the species level is to compare the response curves of related species. For instance, if the optimum value of a curve shifts as in Figure 4.1(*a*) from the last chapter, then species with high levels of resistance to one extreme of an environmental factor should have lower resistance to the opposite extreme of the same factor.

Species comparisons were used by Huey & Kingsolver (1993) to investigate such constraints. As we have previously mentioned, the ability of lizards to sprint may be closely related to their fitness because it influences their ability to capture prey and to escape predators and stressful environmental conditions. Huey & Kingsolver considered sprint speed in lizard species from a number of families under a range of temperature conditions. Data collected from each species enabled the optimum body temperature for sprinting to be determined, as well as the critical high and low temperature extremes when animals were no longer able to sprint. To test for an association between a lizard's optimal body temperature for sprinting and the temperature it experiences in the field, a correlation analysis was carried out. Because each lizard species was not independent of other species, a modified correlation analysis was undertaken after correcting for the phylogenetic relatedness among the lizard species. This analysis provided evidence for an association between the optimal body temperature for sprinting and the average temperatures lizards experienced in the field (Figure 5.8), suggesting that evolution has taken place to enable lizards to perform best in their own environments. In addition, species with highest optimum temperatures for sprint speed could also tolerate the highest temperatures. However, there was no association between the highest and lowest temperatures that the lizards could tolerate.

In terms of the response curves in Figure 4.1 from the previous chapter, these results indicate that the optimum temperature for sprinting differs between species. Because species with the highest optima tolerate the highest temperatures, evolutionary shifts in the optimum value seem to be associated with shifts in the critical maximum in the same direction as in Figure 4.1(*b*). This suggests that temperature optima and resistance to high temperatures are constrained and do not evolve independently. However, resistance to the two temperature extremes seems to evolve independently. Hence there is evidence for only one of the long-term constraints.

We should emphasize that the lizard species used in this study were not reared under identical conditions. As a consequence, differences between species could be a result of acclimation rather than having a genetic basis, even though the large range of optimal temperatures makes genetic differences likely.

An overview of the optimality approach

In this section, we turn to the optimality approach for identifying traits involved in environmental adaptation. If the same characteristics of an

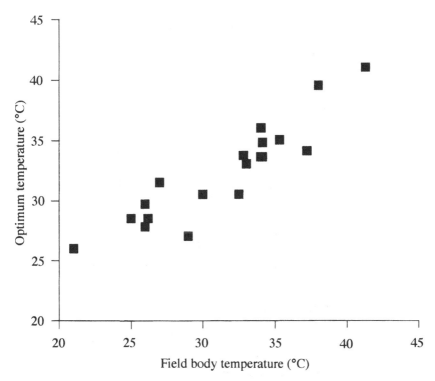

Figure 5.8. Association between the optimum temperature for sprinting in 19 species of iguanid lizards and the average temperature these species experienced in the field. (Simplified from Huey & Kingsolver, 1993).

organism have been favoured by selection for a long time, the common phenotype should represent the 'optimal' phenotype that has evolved because it is adapted to a particular set of ecological conditions. Models can be used to determine which phenotype from a range of possibilities is likely to be optimal. The predicted phenotype can then be compared to the common one found in a population. Optimality models can also be used to examine the relative importance of different factors such as juvenile versus adult survival in producing optimal phenotypes.

It might appear surprising that we are considering this approach in a book about changing conditions. Environments that change are unlikely to result in the evolution of phenotypes that are optimal for particular conditions. Nevertheless, the optimality approach provides insights into how organisms adapt to variability in the environment. We have so far discussed situations where environmental changes are mainly directional, ultimately involving the replacement of one set of conditions by another set, such as when they

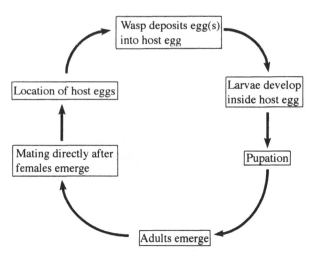

Figure 5.9. Life cycle of a *Trichogramma* wasp which breeds in the eggs of moths.

become wetter, cooler or increasingly contaminated by pollutants. In addition, environments can change with respect to their degree of variability, particularly if extreme conditions are encountered regularly. For instance, temperature or humidity levels may show increasing levels of fluctuations when forests become fragmented into small areas and organisms are less buffered by the presence of a tree canopy. As well as adapting to directional environmental changes, organisms can adapt to increasing (or decreasing) levels of variability. The optimality approach provides insights into how changing levels of variability influence the evolution of traits.

Optimality predictions have been particularly important in life history theory. This body of theory attempts to predict the types of characteristics organisms possess to complete their life cycles, such as size and age at maturity, growth rate, size and number of progeny produced, and patterns of mortality. For instance, the life history of a *Trichogramma* wasp that parasitises the eggs of moth hosts is given in Figure 5.9. The wasp locates the moth eggs, and usually deposits a single egg in each host egg, although several may be deposited in large host eggs. A wasp may parasitise tens or hundreds of eggs in this way within a short lifespan lasting a few days. Females usually produce more female progeny than male progeny. The wasp larva feeds on the moth egg and pupates inside it after several days. The moth egg takes on a darkened appearance prior to the adult emerging. Adult males develop faster than females and mate with females as they emerge. A female attains reproductive maturity within a few hours, and the life cycle continues.

Life history theory attempts to understand why *Trichogramma* have evolved such a life cycle. For instance, life history theory can be used to understand why *Trichogramma* produce a particular ratio of males to females, why wasps deposit a particular number of eggs per host or why wasps live for a short time but mature quickly. To make these predictions, life history theory considers how different characteristics contribute to the lifetime fitness of organisms, as discussed in detail in Stearns (1992), Roff (1992) and Charlesworth (1994).

Before considering the use of optimality approaches to predict life history evolution in variable environments, we need to examine ways of measuring which phenotype or genotype is optimal. This can be straightforward in a simple situation. Consider the case of individuals living in a constant environment and where generations are non-overlapping. The optimal phenotype in such environments is the one resulting in the highest reproductive output within a generation. However, once generations become overlapping, determining what constitutes an optimal phenotype or genotype becomes more complex. This problem is usually approached by considering the rate at which a population grows. A quantity known as the 'instantaneous rate of growth', defined by the symbol r, is often used to describe the rate at which a population expands when all individuals are derived from a single clone or when they have the same phenotype. When population size is not affected by the number of individuals present (ie. population size is not regulated by density), r can be determined from the equation

$$N_t = N_o e^{rt},$$

where N_o is the original population size and t is time. The r value of clones or phenotypes provides a measure of fitness when the size of a population changes because clones with higher values of r will eventually take over the population.

Most populations contain individuals of different ages, and this needs to be accomodated when measuring fitness. We have to consider the survival rate of each age class and the likely contributions that the different age classes make to the next generation. Consider a group of individuals born at a particular time, having a total reproductive life of 7 years. We can follow the progress of these individuals over the 7-year period by measuring their death rate and survival rate. This type of data forms what is known as a 'life table' as in the hypothetical population in Table 5.2. Such information can be used to determine the contribution different phenotypes make to a population. The total reproductive success of the population is given by summing the $l_x m_x$ values, which comes to 1.905 and provides a measure of the 'net reproductive rate' for this age-structured population, a quantity defined by R_0.

Life tables are often represented in a form known as a Leslie matrix. The distribution of age classes in a population at time $t = 1$ can be given by a vector,

$$\mathbf{N}_t = \begin{bmatrix} N_{t,1} \\ N_{t,2} \\ \cdot \\ N_{t,v} \end{bmatrix}$$

Table 5.2. Life table for a group of individuals surviving to a maximum of 7 years

Age class (X)	Number of individuals in age class at start (S_x)	Number of individuals that die within age class (D_x)	Survival rate of age class (p_x)	Survival rate to start of age class (l_x)	Reproductive rate of age class (m_x)	Reproductive success ($l_x m_x$)
0	1000	200	0.80	1.00	0.00	0.00
1	800	200	0.75	0.80	0.00	0.00
2	600	150	0.75	0.60	0.00	0.00
3	450	50	0.89	0.45	1.50	0.625
4	400	100	0.75	0.40	1.50	0.60
5	300	150	0.50	0.30	1.50	0.45
6	150	100	0.33	0.15	1.20	0.18
7	50	50	0.00	.05	1.00	0.05

Note:
Individuals start to breed (i.e., have a non-zero birth rate) after 3 years.

where $N_{1,0}$ is the number at age class 0, $N_{1,2}$ the number at age class 1, and so on up to age class v. The way this vector changes over time describes how the age structure of a population changes. One way this can be obtained is by multiplying the vector by a square matrix known as a 'Leslie matrix'. One form of this matrix is defined as

$$
\mathbf{L} = \begin{bmatrix}
0, & F_1, & F_2 & . & . & F_v \\
p_1, & 0, & 0, & . & . & 0 \\
0, & p_2, & 0, & . & . & 0 \\
0, & 0, & p_3, & . & . & 0 \\
0, & 0, & 0, & . & . & p_v
\end{bmatrix}
$$

where $F_1, F_2, ... F_v$ are the age-specific fecundities of females, and $p_1, p_2 ... p_v$ are the survival rates. In this form, individuals from the life table are sampled after breeding has occurred. Values in this matrix can be computed from data such as those in Table 5.2. Fecundities are obtained by multiplying the birth rates, m_x, by the survival rate of the previous age class, p_{x-1}. In our example, the Leslie matrix is given by

$$
\begin{bmatrix}
0 & 0 & 1.125 & 1.335 & 1.125 & 0.6 & 0.33 & 0 \\
0.8 & 0 & 0 & 0 & 0 & 0 & 0 & 0 \\
0 & 0.75 & 0 & 0 & 0 & 0 & 0 & 0 \\
0 & 0 & 0.75 & 0 & 0 & 0 & 0 & 0 \\
0 & 0 & 0 & 0.89 & 0 & 0 & 0 & 0 \\
0 & 0 & 0 & 0 & 0.75 & 0 & 0 & 0 \\
0 & 0 & 0 & 0 & 0 & 0.5 & 0 & 0 \\
0 & 0 & 0 & 0 & 0 & 0 & 0.33 & 0
\end{bmatrix}
$$

To work out how the vector of age classes changes from time t to time $t+1$, we can then compute $\mathbf{N}_{t+1} = \mathbf{N}_t \mathbf{L}$. Matrices can also be constructed for organisms that undergo a series of discrete life cycle stages as described in Stearns (1992).

In an age-structured population, the instantaneous rate of increase (r) can be obtained when taking the different age classes into account, by solving the equation

$$
\sum_{X=g}^{X=w} e^{-rX} l_x m_x = 1,
$$

where w is the age at last reproduction and g is the age at maturity. If r is zero because the population is stable, this equation reduces to

$$
\sum_{X=g}^{X=w} l_x m_x = 1.
$$

Under these conditions, the net reproductive rate for the entire population (R_0) is equal to one because on average only one offspring is produced by each female.

By measuring R_0 for individuals with different characteristics (such as size, development time or maturation time) from a stable population, an estimate of lifetime fitness associated with these characteristics can be obtained. However, this measure is not useful when population size is changing, because R_0 only considers the number of progeny produced by individuals, and not the timing of reproduction. For instance, if a population is expanding, individuals with genotypes that reproduce relatively early will be favoured because they produce progeny earlier, increasing the representation of this genotype in the population.

Once environmental variability is introduced, the estimation of fitness becomes more complex because fitness estimates need to be averaged across environments. As we mentioned in the first chapter, the mean fitness (w) of a genotype is usually represented by its geometric mean fitness when environments fluctuate rather than its arithmetic mean fitness (Cohen, 1966; Gillespie, 1973), i.e.

$$\overline{w} = {}^N\sqrt{X_1 X_2 X_3 \cdots X_N}$$

rather than

$$\overline{w} = \frac{X_1 + X_2 + X_3 \cdots X_N}{N},$$

where X_1, X_2 ... X_N are the fitness values of the phenotypes in the N environments. As a consequence, averaging fitness measures such as r will overestimate the overall fitness across environments. The fitness of a genotype or phenotype in unfavourable conditions can have a particularly large influence on its overall fitness. For instance, consider two genotypes in five environments, one of which is unfavourable. The fitness of the first genotype is not affected by unfavourable conditions, but it only has a fitness of 0.6 in all environments, whereas the second genotype normally has a fitness of 0.9, but a low fitness (0.05) in the unfavourable environment. On the basis of the arithmetic mean, the first genotype has a lower overall fitness (0.6) than the other genotype (0.73). However, the geometric mean fitness of the first genotype (0.6) is higher than that of the second genotype (0.5) even though it performs more poorly in four of the five environments. The optimal phenotype likely to evolve in a population can therefore be difficult to predict unless there is information on relative fitness under extreme as well as favourable conditions.

The geometric mean is always appropriate when a population experiences environmental variability (Bulmer, 1985), but computing overall fitness using the arithmetic mean may be appropriate when there is migration among different environments (Hastings & Caswell, 1979). In this case, a phenotype with a high fitness in a good environment but zero fitness in a bad environment could still persist, because some individuals with this phenotype evade stressful conditions. If there is efficient dispersal and good environments are common, this phenotype may be favoured most of the time because it has a high arithmetic fitness. In general, migration will occur from good (source) populations to bad (sink) populations, and fitness in good environments will be of overriding importance (Houston & McNamara, 1992; Kawecki & Stearns, 1993).

To measure the fitness of a genotype over a long interval, its contribution to the long-term growth rate of a population in a variable environment needs to be considered. Following Orzack & Tuljapurkar (1989), the population growth of a population occupying an environment that changes in time is given by

$$N_{t+1} = L_{t+1} N_t = L_{t+1} L_t ... L_1 N_0$$

where L_t is the Leslie matrix at time t and N_0 is the initial vector of age classes. Values in the Leslie matrix may change randomly as populations encounter a random set of environments. In this situation, the growth rate of an age-structured population can be shown (Tuljapurkar, 1982; Metz *et al.*, 1992) to equate with

$$a = E \frac{1}{t} [\ln M_t - \ln M_0]$$

where M_0 and M_t represent the population sizes at times 0 and t, and E denotes an arithmetic expectation. The a term is the stochastic or long-term growth rate of the population, and is also known as the dominant Lyapunov exponent (Metz *et al.*, 1992). Values of a can be greater or less than zero, depending on whether a population is increasing or decreasing. This parameter is the appropriate measure of growth rate in all situations including unpredictable and variable environments. Although estimates of a are difficult to obtain analytically, they can be obtained numerically as long as long-term data are available from natural populations. For instance, estimates of a have been obtained by Benton *et al.* (1995) for a population of red deer based on 21 years of data.

As well as describing growth rate, a provides a measure of fitness in a variable environment, analogous to r or R_0 in a constant environment. This measure is important in predicting the evolution of life histories in variable environments. Consider a simple genetic situation where a life history pattern is determined by one gene with two alleles, generating three genotypes, A_1A_1, A_1A_2 and A_2A_2. If a population consists solely of A_1A_1 individuals, then the ability of individuals carrying the A_2 allele depends on their stochastic growth rates. In other words, the A_2 allele will increase in frequency if the growth rate of the A_1A_2 genotype exceeds that of the A_1A_1 genotype, or $a_{A1A2} > a_{A1A1}$.

In summary, three measures of fitness used in life history analyses are r, R_0 and a. The first two measures approximate fitness when populations are in relatively constant environments. The last measure was devised specifically to address fitness in variable environments. These measures can be used to predict which genotype or phenotype will become predominant in a population.

To consider the effects of environmental variability on the optimal life history patterns likely to evolve, some theoretical models use the ESS (Evolutionary Stable Strategy) approach (Maynard Smith, 1982). The optimal phenotype is determined by considering a series of alternative phenotypes, and determining which one has the highest fitness when present in combination with the other phenotypes. ESS approaches start with setting up a 'payoff' matrix, which describes the benefits an individual obtains from having a particular strategy in a particular environment. As an example, consider the following strategies, A and B, that describe the dispersal behaviour of individuals. While A individuals do not disperse, two offspring from B individuals disperse to another environment. We can examine the payoffs of these strategies in good and bad years with the shown matrix:

	Good year	Bad year
Strategy A	N	Ns_1
Strategy B	N-$2s_2$	N-$2s_2$

In a good year, strategy A results in N offspring that all survive, but in a bad year, only a fraction (s_1) survive. The B individuals also produce N offspring, but two of these disperse, and the total number of progeny surviving depends on the probability of these dispersers reaching a suitable environment and surviving (s_2). In a bad year, all offspring produced by B survive because there are two fewer, increasing the amount of resources available for each offspring. This payoff matrix can be used to examine which strategy has the highest fitness (ie, produces the most offspring). If the frequency of good environments is given by k, then the productivity of A is given by $kN+(1-k)(Ns_1)$, while that of B will be given by $k(N$-$2s_2)+(1-k)(N$-$2s_2)$. These 'payoffs' can be used to examine the effects of varying frequencies of good and bad years on the expected fitness of the strategies. For instance, if N is set at 5 offspring, and s_1, s_2 at 0.1 and 0.5 respectively, the strategies will have the same fitness when $k=0.78$. Strategy A will be favoured when good environments occur more frequently than this value, and B will be favoured when they are less frequent. Models of this type can therefore indicate the level of dispersal expected under variable conditions.

Incorporating extreme environments into optimality predictions

The above fitness measures and approaches can help to investigate the impact of environmental extremes on life history evolution. We will see that the optimal traits predicted by theory may change markedly when unfavourable conditions are included in models. However, empirical data to test such predictions are relatively scarce.

1. Seed dormancy and diapause

Seeds will often germinate with the onset of rainfall and other conditions. If favourable conditions can be predicted to occur normally, there is little need for plants to maintain a seed bank by having seeds with different levels of dormancy. However, good growth conditions may be unpredictable. For instance, in deserts rainfall can be sporadic and often insufficient to support the development of a plant.

When the environment is constant, complete germination without dormancy is always favoured because seeds that germinate early will produce more offspring than those with delayed germination. However, if unfavourable years occur sporadically, only dormant seeds will survive on some occasions. This can have drastic effects on the predicted life history of a plant, particularly in annuals, where a single bad year could lead to the extinction of an entire population. Obviously, some degree of dormancy will always be favoured in such plants, even if stressful years are extremely rare. Seed dormancy may therefore be an example of a 'bet-hedging' strategy. By producing some dormant seeds as well as non-dormant seeds, plants are maximising the chances of some seeds surviving extreme conditions; ie., they are hedging their bets by producing some seeds that have a high fitness under stressful conditions and others with a high fitness under favourable conditions. This is an example of 'diversified' bet hedging because plants maximise fitness by producing diverse offspring.

Cohen (1966) considered the exact proportion of seeds that were expected to be dormant. He showed that when there are only good and bad years, and when the seed yield in bad years is 0, then the proportion of dormant seeds produced by plants will be reduced as the number of good years increases. Theoretical models also predict that a constant fraction of seeds remaining after the first year germinate in subsequent years. These predictions are based on the population's growth rate being determined by the geometric mean fitness across good and bad years.

Although the predictions from models seem clear-cut, there are difficulties in relating them to empirical data (Philippi, 1993). Dormancy can be imposed by environmental factors rather than representing the outcome of evolution for bet-hedging. For instance, dormancy may be imposed by deep burial. True dormancy only occurs if seeds fail to hatch in the first year regardless of environmental conditions, although it is also possible that the degree to which seeds respond to environmental conditions is under selection. In desert annuals, Philippi (1993) showed that some seeds fail to germinate in their first year despite favourable conditions, indicating true dormancy. However, in contrast to theoretical predictions, the germination percentage in subsequent years varied between species and did not constitute a constant fraction of the remaining seeds. This suggests that seed germination patterns may not closely follow model predictions and that factors that have not yet been defined are involved.

Diapause in insects may represent another case of diversified bet-hedging. In some insect populations, many individuals emerge from diapause early, whereas others stay in diapause for a long time. For instance, in cabbage root flies, emergence may occur after 14 days or after more than 100 days (Finch & Collier, 1983). This could reflect responses to environmental conditions because different temperature and light regimens induce diapause in some individuals but not in others (Tauber *et al.*, 1986). Because diapausing stages have a high level of stress resistance, individuals leaving diapause late evade stressful conditions at the beginning of a season. In contrast, early emergence will be favoured if stressful conditions are not encountered at this time.

At this stage, it is not clear if individual differences in diapause characteristics tend to reflect bet-hedging. If individuals in a population follow the same optimal strategy, all are expected to produce diverse progeny, some of which undergo diapause and others do not. However, individual differences in the tendency to diapause can often be attributed to genetic factors. Populations of milkweed bugs and many other insects contain individuals that differ genetically in their tendency to diapause (eg Dingle, 1981). These individuals produce progeny with the same diapausing tendency as themselves rather than diverse progeny.

2. Timing of reproductive effort

Organisms may reproduce once or many times in their life time, which are life history patterns known respectively as semelparity and iteroparity. Cole (1954) set up a model comparing the fitness of an organism reproducing once and living for a year (such as an annual plant) with that of a perennial living forever, where fitness was defined as the intrinsic growth rate of a population. Under the assumption that there was no juvenile mortality, he concluded that semelparity is advantageous over iteroparity in a constant environment, a result that has become known as Cole's paradox. This paradox can be overcome by introducing a tradeoff between present reproduction and future reproduction. If organisms put all their resources into reproduction, there will no longer be any resources left for future survival and reproduction.

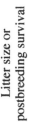

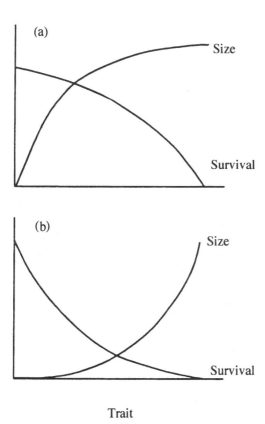

Trait

Figure 5.10. Concave and convex associations between two measures of fitness (litter size, survival) and variation in a trait. When fitness functions are concave (a), size increases relatively faster (and survival declines relatively slower) at low trait values. Under convex fitness functions (b), size increases relatively faster (and survival declines relatively slower) at high trait values.

On the other hand, the continued survival and reproduction of an iteroparous pattern can only occur if fewer resources are devoted to present reproduction. There is evidence for such a tradeoff. For instance, Young (1990) listed studies showing that semelparous plants have a 1.5–5 fold higher reproductive output per episode of reproduction than iteroparous plants.

The effect of this tradeoff on the timing of reproduction depends on how changes an organism makes in its reproductive effort affects its survival and the size of the brood it produces (Schaffer, 1974). This association may be 'concave' or 'convex'. In the former case, the survival of adults decreases slowly with a trait at first, and later decreases more rapidly, producing a graph that has a concave shape (Figure 5.10). In contrast, the number of progeny increases rapidly initially, then more slowly. Under these conditions, the reproductive success of an organism is at a maximum at intermediate levels of reproductive effort. The optimal strategy involves survival for some time without the production of a maximum brood size (iteroparity). On the other hand, convex fitness functions lead to the opposite result because fitness is at a maximum when reproductive effort peaks. This selects for a maximum brood size at the expense of adult survival, favouring semelparity.

What are the effects of changing environmental conditions on these predictions? As in the case of dormancy, environmental variability can be introduced by assuming that there are only good and bad years, and that overall fitness is determined by the geometric mean across years. Schaffer (1974) showed that when changes in the environment influence adult survival, then environmental variability will favour an annual over a perennial as long as reproductive output is relatively high. He also showed that perennials will be favoured when environmental changes influence juvenile survival. These results make intuitive sense; when bad years reduce juvenile survival, an iteroparous pattern will be favoured because it allows the risk of reproduction to be spread over several years, like the bet-hedging strategy discussed above in the context of seed dormancy. On the other hand if adult survival varies, there is not much to be gained from postponing reproduction.

Predictions can also depend on the age structure of a population and the way environments change. Orzack & Tuljapurkar (1989) considered the effects of age structure and environmental variability on reproductive patterns by using an approximation of the stochastic growth rate (a) discussed earlier as a measure of fitness. This measure leads to complex predictions about the optimal pattern, depending both on the degree to which environments are variable, and the extent to which growth rates in environments are correlated. The latter addresses how environmental conditions vary in time or space. Conditions may vary randomly, in that a poor year with low growth rate is just as likely to be followed by another poor year as by a favourable year. On the other hand, poor years may tend to occur in clusters as in the case of El Niño years, in which case there will be a positive correlation in growth rates across environments. Poor years might also tend to be followed by favourable years, producing a negative correlation in growth rates across environments.

An example of the type of results obtained by Orzack & Tuljapurkar (1989) is presented in Figure 5.11, and considers the stochastic growth rates of four life history patterns when growth rates are negatively correlated across environments. Fitness is plotted against the level of variability in the environment. When environments are not variable, the favoured pattern involves reproduction over a short interval and a short generation time (strategy A). This matches Cole's result discussed above. As variability increases, a longer generation time is favoured even when the reproductive interval is the same. At higher levels of environmental variability, individuals that reproduce over

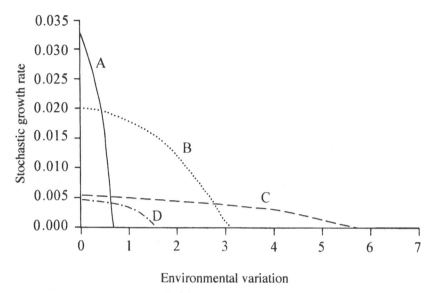

Figure 5.11. The stochastic growth rate of four life history strategies when growth rates across environments are negatively correlated ($r = -0.30$), and environments show different levels of variability. Strategy A reproduces at an early age (say day 1) for a short period (days 1–2). Strategy B also reproduces on day 1 but for a longer time (1–4 days), and has a lower fertility per day compared to A. Strategy C reproduces later (day 7) until day 10 and has the same fertility as B. Finally, strategy D has the same fertility as A but only reproduces on days 9–10. (Simplified from Tuljapurkar & Orzack, 1989).

longer intervals have the highest growth rates, favouring iteroparity (strategy B, C). Individuals that delay their reproduction (C) can do particularly well in variable environments. When growth rates in different environments are positively correlated, these individuals may be at an advantage even when they reproduce over a short period. Variability within the environment can therefore lead to complex predictions. Such effects have recently been confirmed by Benton and Grant (1996) who used numerical estimates of a to predict life histories. They showed that the types of life histories organisms evolved were affected by environmental conditions, even when populations rarely encountered stressful conditions.

These results differ to some extent from those obtained by Schaffer (1974). Nevertheless, models that include age structure often indicate an advantage of spreading reproduction when increasing environmental variability influences fecundity and juvenile survival rate (see also Goodman, 1984).

Few attempts have been made to test predictions about environmental effects on life history patterns. Murphy (1968) compared the association between reproductive span and variation in spawning success in five schooling fish living on plankton. When spawning success was used as a measure of environmental uncertainty, there was a strong positive relationship between these variables, suggesting that environmental variability was associated with an increased reproductive span. However, Roff (1992) suggested that one of the data points used by Murphy was incorrect, and his alternative value reduced the strength of the correlation. He also provides a fish example that did not fit this prediction.

Young (1990) examined the effects of environmental variability on reproductive success in two plant species from the genus *Lobelia*. He monitored populations of the species from Mount Kenya over an eight year period. One of the species, *Lobelia telekii*, occurs on well-drained soils and is semelparous, producing a single rosette. The other species, *L. keniensis*, is iteroparous, occurs in wetter sites where there is more vegetation, and produces branches with several rosettes. Both species take 40–60 years to reach maturity, and the semelparous species is 3–5 times more fecund than the iteroparous species. Young (1990) found an association between variability in soil moisture and reproductive output. In both species, the number of seeds set increased under moist conditions. In addition, the intervening period between episodes when the plants set seed was reduced in moist conditions. This led to a higher mortality of adult *L. keniensis* between reproductive episodes when the soil was dry. This information was used to predict which life history patterns are expected in lobelias. On the basis of an optimality model where there was no population growth (ie $R_0=1$) , Young found that the 3–5 fold reproductive advantage of the semelparous species was sufficient to explain semelparity in dry sites where adult survival was reduced. However, in wet sites where adult plants have an increased chance of survival, a semelparous individual would have to be >13 times more fecund to be at an advantage over an iteroparous plant. This suggests that semelparity has been favoured when dry conditions result in high levels of adult mortality between episodes of reproduction.

This example illustrates how predictions from specific models can be tested when there is information on environmental conditions and their effects on survival and reproduction. Such information can only be obtained in long-term studies of natural populations.

3. Predicting offspring number

Closely tied to the issue of iteroparity and semelparity is the question of the number of offspring organisms should produce. To maximise geometric mean fitness, individuals exposed to variable environments may rear fewer progeny than the maximum number they can rear under favourable conditions. Following Philippi & Seger (1989), we consider three possible phenotypes, designated as A, B and C. These phenotypes are associated with

Table 5.3. *Relative fitness of four phenotypes in a good and bad environment*

	Phenotype		
	A	B	C
good environment	1.00	0.60	0.785
bad environment	0.58	1.00	0.785
arithmetic mean	0.79	0.80	0.785
geometric mean	0.762	0.775	0.785

Note:
The mean fitness of the phenotypes is determined by assuming that the good and bad environments occur equally frequently. Phenotypes A and B are specialists in good and bad environments respectively, having the maximum fitness (1.00) in these environments. C is a conservative bet-hedger.
Source: Modified from Philippi & Seger (1989).

different fitness values in good and bad environments as illustrated in Table 5.3. The A phenotype produces the largest number of offspring. It has highest fitness in the good environment when all its offspring can be reared successfully, but a low fitness in a poor environment because few of its offspring survive. In contrast, B produces only a few offspring and all of these survive in the bad environment. Hence only A would be favoured when conditions are good, whereas B would be favoured when they are bad.

But what happens when environments fluctuate, and both good and bad environments occur? In this case, the geometric mean fitness of the phenotypes differs from the arithmetic mean fitness when good and bad environments occur equally frequently. While A and B have a similar overall fitness, phenotype C, which produces an intermediate number of offspring, has a higher fitness than either A or B. Phenotype C has an intermediate fitness in good and bad environments, and will be favoured over A and B even though it does not do best in either type of environment. C is known as a 'conservative bet-hedger' because it does not do best in A or B but does fairly well in both environments.

Such arguments have been used to account for the mean number of offspring produced by organisms living in variable environments. Boyce & Perrins (1987) considered the case of clutch size in a population of the great tit, *Parus major*. Some years were favourable for the survival of young, enabling birds to produce large clutches, whereas other years were unfavourable, enabling only small clutches to survive. The mean clutch size produced by birds in this population was smaller than the clutch size that was successfully reared in favourable years. Although birds producing clutches larger than the mean size would be at an advantage under favourable conditions, they would be expected to rear fewer young in unfavourable years. Boyce & Perrins (1987) computed the geometric fitness of birds with different clutch sizes over a number of years, and found that this was highest for individuals with intermediate clutches. This suggests that clutch size in great tits is an example of conservative bet-hedging.

However, a reanalysis of these data by Liou *et al.* (1993) highlights the types of problems that can be encountered when attempting to fit field data to predictions. They suggested that the observed difference between the mean number of eggs and maximum number that can be successfully reared may reflect the condition of birds rather than an adaptive response to environmental variability. When females are in good condition, they tend to lay eggs earlier. This factor correlated with clutch size in great tits, suggesting that the condition of birds directly determined clutch size irrespective of environmental variability. Liou *et al.* (1993) also showed that geometric means can be difficult to estimate in natural populations.

One reason why environmental variability might not have a major effect on the clutch size of birds is that environmental effects on clutch size are not particularly large. Cooch & Ricklefs (1994) considered the effects of environmentally-induced variability in clutch size and survival on the optimal clutch size expected to be produced by birds. This involved varying clutch size by up to 20%, and survival by up to 15%. The variability had little effect on the optimal reproductive effort in simulated populations, causing shifts in clutch size of only a few per cent. The authors suggested that differences in reproductive effort among bird populations may be related to the availability of food, rather than representing an optimal response to environmental variability.

4. Hatching asynchrony

Apart from clutch size, the way eggs hatch may also be affected by environmental variability. Lack (1954) proposed that, in birds, synchrony in hatch rates influences competition among hatchlings in a brood. Young hatching late will be at a disadvantage in competition with their older siblings. This effect may be small when food is abundant, but the youngest nestlings can die when food is scarce. In the latter situation, a parent's reproductive success will be increased because the death of some young enables parents to rear the others successfully, whereas all young would otherwise die in a poor year.

This hypothesis has been treated theoretically by Pijanowski (1992), using an ESS approach. The matrix considers payoffs in good and bad years and is given as shown.

	Good year	Bad year
Synchronous	B	$s.B$
Asynchronous	$B-1+q$	$B-1$

In the matrix, B represents the brood size and s represents the expected survival of a synchronous hatching, varying between 0 and 1.0. In a good year, all brood of a synchronous strategy survive, whereas the payoff is $s.B$ in a bad year. In the asynchronous strategy, one individual hatches later than the others. This individual does not survive a bad year whereas the others do (payoff $B-1$). In a good year, the survival rate of this individual is given by q so that payoffs are equivalent in good years if all progeny survive.

Using this matrix and allowing nestling survival to decrease exponentially with brood size, Pijanowski (1992) examined the effects of different frequencies of good and bad years. As expected, a decreasing frequency of good years favoured hatching asynchrony. This strategy remained advantageous even when there was reduced survival associated with asynchrony in a good year (ie., $q<1.0$). However, synchronous hatching became advantageous as the effects of bad years became more drastic. This is because most progeny in this situation are produced from good years when synchronous hatching is at an advantage (Lamey & Lamey, 1994). These results are summarized by Figure 5.12 which shows that as the frequency of good years increases, and bad years become more severe, so does the area of parameter space when synchrony is favoured.

Bird data have demonstrated that last-hatched nestlings do not survive as well as their siblings in good years (Amundsen & Stokland, 1988), so q is generally less than 1. This cost needs to be offset by lower survival of sychronous nestlings in bad years. There is some evidence that q can be high enough to favour asynchrony. For instance, Magrath (1989) manipulated both food availability and asynchrony in blackbirds (*Turdus merula*). Variation in hatching pattern was mimicked by swapping recently-hatched nestlings between nests, while food availability was manipulated by providing a supplement to some pairs of blackbirds. Differences between asynchronous and synchronous breeds were consistent with expectations (Figure 5.13), particularly in bad years when asynchronous broods showed an increased fitness.

The advantage associated with asynchrony emphasizes that measurements of fitness in one environment may lead to erroneous conclusions about the relative fitness of different phenotypes. As pointed out by Pijanowski (1992), asynchrony can still be favoured even though this strategy results in fewer progeny being raised in a good year. However, we should emphasize that there are explanations for asynchronous patterns that are unrelated to environmental variability as in the case of clutch size.

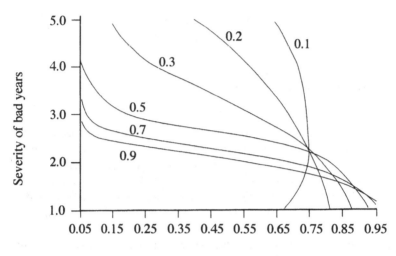

Figure 5.12. Predicted success of broods that hatch asynchronously versus synchronously. Lines separate conditions when synchronous hatching is favoured (above and right of line) over asynchronous hatching (left and below line). Lines are given for different values of a constant reflecting how nestling survival and brood size change. Small values of the constant indicate increased nestling survival with brood size. (From Lamey & Lamey, 1994).

5. Dispersal ability

Dispersal enables organisms to locate favourable areas and alter the level of environmental variability they experience. Dispersal also allows organisms to evade crowded conditions. Like dormancy, dispersal can result in escape from extreme conditions, and increasing levels of environmental variability are expected to select for an increased rate of dispersal.

Theoretical models of dispersal have focussed on plants because these cannot move in response to poor conditions, simplifying the models. An example of such a model is the ESS approach used by Levin *et al.* (1984) to consider an annual population distributed over a series of patches varying in quality. In their model, patches with good conditions resulted in more seeds than those with poor conditions, and populations were influenced by density dependence in each patch so that fitness declined as a consequence of crowding. Non-dispersers were at an advantage if a poor patch became good in the following year. However, because of crowding, dispersers were at an advantage if good conditions persisted in the following year. Hence there is an advantage and a cost to dispersal, and its magnitude depends on the number of dispersers in a population. The optimal strategy in this case was an inter-

Good conditions

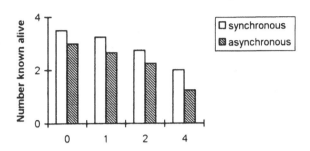

Poor conditions

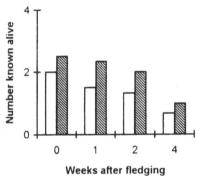

Weeks after fledging

Figure 5.13. Survival of nestlings in broods of blackbirds reared under good and poor conditions where hatching occurred synchronously or asynchronously. Bars represent the mean number of nestlings that were fledged and survived 0–4 weeks after fledging. (From Magrath, 1989).

mediate level of dispersal, depending on the level of environmental variability; more variable conditions selected for increasing levels of dispersal.

The optimal level of dispersal will also depend on the way environments vary in space and time and on the age structure of a population (Levin & Cohen, 1991; Wiener & Tuljapurkar, 1994). If good environments tend to occur together more often than expected on the basis of randomness, variability between adjacent environments will be reduced. This tends to select for a reduction in the dispersal fraction of a population. On the other hand, if good conditions tend to be associated with bad conditions, dispersal is favoured to a greater extent. Wiener & Tuljapurkar (1994) considered the effects of migration on the long-term growth rate of an age-structured population by defining fitness in terms of a, the stochastic growth rate. They found that the presence of age structure reduced the benefits of dispersal in a variable environment. As a consequence, selection for dispersal under a range of conditions became weak and a range of dispersal values had the same fitness.

Models of dispersal have not been related closely to empirical data. There is little doubt that high rates of dispersal often occur in organisms exposed to variable environments. For instance, in terrestrial insects, the predictability of habitats is closely associated with their tendency to migrate (Southwood, 1962). However, other predictions from dispersal models remain to be tested, such as whether dispersal is reduced in populations with age structure and in environments where there is a high incidence of poor years.

Optimality predictions and trait interactions

We have so far considered the effects of environmental variability on optimality predictions for individual traits. However, there will often be interactions among traits that constrain the types of evolutionary change that can take place. For instance, dormancy, dispersal, adult longevity and seed size are not expected to vary independently in plants but to place constraints on one another. Light seeds are expected to disperse further than heavy seeds, and are less likely to show dormancy because they have insufficient energy reserves to emerge when they become buried.

Such interactions can be included in models to determine how traits evolve together in variable environments. One approach to this problem is to consider how the effects of environmental variability on one trait influences the evolution of another trait. For instance, Rees (1994) used an ESS approach to investigate the interaction between survival, dormancy and longevity from this perspective, by considering which dormancy strategy would be an ESS under different sets of conditions. He found that when environmental variability influences the survival of seedlings, increases in longevity select against seed dormancy. This fits intuitive expectations because long-lived plants will persist during unfavourable periods and produce seeds to colonize new areas, whereas short-lived plants may die within unfavourable periods. However, such predictions became counterintuitive when the effects of environmental variability on the survival of adults as well as seedlings was considered. In this case, plants with intermediate longevities were predicted to have the lowest levels of dormancy.

The intuitive prediction has been related to empirical data. Rees (1993) examined seed dormancy in 171 species from 34 families ranging from annual herbs to perennial grasses. He used the fraction of seedlings emerging after one year as a measure of dormancy. Phylogenetic information enabled relatedness among species to be controlled by setting up a series of independent contrasts as described earlier in this chapter. After correcting for seed size, Rees found that long-lived plants showed less dormancy as expected (Figure

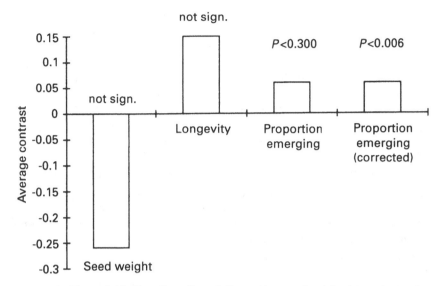

Figure 5.14. The effect of seed dispersal on seed weight, longevity, and the proportion of seeds emerging in the first year, after corrections are made for phylogenetic relatedness among species. A positive contrast indicates that efficient seed dispersal is associated with an increase in the character, and conversely for a negative contrast. The corrected bar refers to the contrast after the effects of seed size and longevity on dormancy have been corrected. Probabilities indicate if the trends are significantly different from zero (no association). (Simplified from Rees, 1993).

5.14). This trend was apparent in comparisons of annuals and perennials from both the European and North American flora.

In insects, interactions may occur between size, dispersal and development time. Large insects generally fly further than small insects (Roff, 1991), so migrants are expected to be large compared to non-migrants. In an uncertain environment where dispersal is favoured, this can lead to selection for a large body size. However, insects may have difficulty attaining a large size because it can be difficult for larvae and pupae to complete development in these uncertain environments. To overcome this problem, insects must develop quickly, either by developing into an adult early or by growing rapidly. Roff (1991) argued that the latter is more likely in migrants, because rapid development can result in a small body size and decreased migratory ability. Roff presented comparative data to support this idea. For instance, in butterflies, species that migrate because of changing environmental conditions tend to be larger and have faster growth rates than non-migrants. The effects of body size on dispersal ability may therefore help to determine the way insects develop in uncertain conditions.

We have so far considered the effects of interactions on the means of traits. Interactions influence the expected mean level of dormancy level of seeds or growth rate of migrants. A different approach is to consider how interactions between traits influence the way the traits themselves are altered under changing conditions. In other words, given particular interactions among traits, which plastic changes organisms make in response to environmental variability are optimal? Stearns & Koella (1986) used this approach when considering how environmental effects on growth rate are expected to affect the age and size organisms reach at maturity. To measure fitness, they used r, the instantaneous rate of increase of a population. They also assumed that fecundity increases with body size and age, and that the survival of juveniles and size at maturity are increased if maturity is delayed. This results in a 'conflict' when trying to maximise r; a short generation time increases r, but this leads to decreased reproductive output and juvenile survival which decrease r.

Because of the number of variables in the model, predictions were complex and resulted in diverse responses to changing conditions. Slow growing individuals were expected to mature at a relatively large size under some conditions, but at a relatively small size under different conditions. Predictions depended on how slow growth influenced juvenile and adult mortality levels. Moreover, Berrigan & Koella (1994) showed that model predictions also depend on how fitness is maximised. When the net reproductive rate of a population is maximised (the R_0 term mentioned earlier in this chapter) rather than r, the way growth rate influenced age and size at maturity changed markedly.

Attempts have been made to relate these models to data. For instance, Stearns & Koella (1986) compared their predictions with the effects of crowding and nutrition on age at maturity in *Drosophila*. When flies are reared under conditions that are crowded for larvae or where nutrition is poor, they emerge as adults at a reduced size, which presumably reflects a plastic response to maximise fitness. This observation is in qualitative agreement with predictions from optimality models where juvenile mortality increases as growth rate decreases. However, when comparisons are made at the quantitative level, Berrigan & Koella (1994) showed that R_0 can also predict the way growth rate affects age and size at maturity (Figure 5.15). The way flies change their age at maturity when environmental conditions reduce growth rate can therefore be predicted given one set of assumptions, although other models with different assumptions will also fit these data.

Several factors need to be considered when applying optimality models to plastic changes in response to environmental extremes. First, predictions based on plasticity models may not apply to all environmental factors that limit growth rate. For instance, while *Drosophila* mature at a smaller size when growth is delayed by crowding and poor nutrition, the opposite may

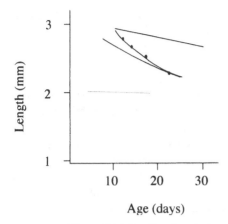

Figure 5.15. Predicted and observed associations between the age and size at maturity of *Drosophila mercatorum* flies. Circles represent data points from laboratory-reared flies. Solid lines represent predictions of reaction norms from three models where R_o is the fitness measure. The dotted line represents the predicted norm when r is used as a fitness measure. (From Berrigan & Koella, 1994).

occur with temperature effects (Berrigan & Charnov, 1994). In insects, high temperatures that increase development rates result in adults emerging at a smaller size, whereas low temperatures slowing development result in larger adults. This trend is widespread in ectotherms (Atkinson, 1994). The reasons why individuals mature at a small size at high temperatures may be unrelated to the conflict between minimising generation time and maximising reproductive output.

Secondly, the environmental conditions themselves may influence the fitness of individuals in ways that are unrelated to age at maturity or development time. Under life history theory, a *Drosophila* reared under crowded or cold conditions and pupating at a particular size is assumed to have the same reproductive potential as an individual pupating at the identical size but reared under uncrowded or warm conditions. There are two problems with this assumption. First, the rearing conditions experienced by individuals may have diverse effects on adult fitness, irrespective of size. For instance, Zamudio *et al.* (1995) showed that *Drosophila* males reared at 25°C had a greater territorial success than males reared at 18°C, even though larger flies normally win territorial contests. Second, the relative fitness of individuals exposed to different rearing conditions will depend on the conditions they experience at the adult stage. This can be illustrated by considering acclimation. When juveniles experience a sub-lethal stress, they often become more resistant to the same stress at the adult stage. Individuals reared under cold conditions will have a relatively higher fitness if they experience cold stress as adults, and vice versa for individuals reared under warm conditions. Predicting how the rearing environment affects fitness becomes difficult if effects on adult fitness are specific to certain conditions.

In conclusion, optimality models of plastic changes are useful in emphasizing that responses to environmental changes can be adaptive, and the models help to spell out the assumptions that need to be tested. However, the models suffer from having too many parameters, making it difficult to relate predictions to empirical data. It appears that many outcomes are theoretically possible, depending on fitness measures and assumptions about associations between growth rates, mortality levels and size. In addition, other factors need to be considered when making predictions, such as the way conditions at one point of an individual's life will influence fitness at a later stage.

Limitations

The comparative and optimality approaches help us to understand how extreme conditions can influence particular evolutionary outcomes. However, these approaches only provide limited insights into how evolution occurs. In addition, they have other limitations which are briefly discussed below.

One limiting factor for the comparative approach is the absence of information about genetic differences between species. As we have already mentioned, variation needs to be genetic if it reflects evolutionary changes. To establish genetic differences between two groups, we should ideally carry out crosses between them to examine patterns of inheritance. Because crosses cannot usually be carried out between species, we need to minimise the chances that differences between them are environmental rather than genetic. This can be done by rearing and testing species under the same conditions. If not, it has to be assumed that the expression of traits is not influenced by the environmental conditions experienced by species. Unfortunately, the expression of most traits is influenced by the environment, often in dramatic ways. For instance, seedlings of the same plant may develop into a tree in one environment but into a shrub in another environment. Rearing individuals in the same environment can be an onerous task for species that are not closely related, because these will usually require different environmental conditions to develop successfully.

The problem of environmental control can also be important when testing optimality predictions. As mentioned above, optimality theory predicts that diversified bet-hedgers will be favoured in populations from variable environments. This means that individuals are expected to produce progeny that have different phenotypes, some adapted to favourable conditions and others adapted to stressful conditions. The production of different phenotypes by the same individual is often inferred by individual variation in a population, such as the presence of dormant and non-dormant seeds in plant populations. However, such differences can also be determined by genetic variation or by the environment.

The effects of environmental conditions on species differences are further confounded by what are known as 'cross-generation' effects. It is becoming increasingly apparent that the environments parents experience will often influence the phenotypes of their progeny. For instance, the length of time insects spend in diapause is often influenced by the number of daylight hours experienced by their parents (Mousseau & Dingle, 1991). As a consequence, under ideal circumstances, both parents and progeny need to be reared in the same environment when comparisons are made between species or when an optimal phenotype is being identified.

A second factor limiting the usefulness of both the comparative and optimality approaches is the problem of genetic interactions among traits. If two traits are affected by the same genes because of the phenomenon of pleiotropy as discussed in Chapter 1, evolutionary changes in one trait can result in changes in the second trait, even when the second trait is not under selection. As a consequence, spurious correlations between environmental variation and the other trait can occur. Because traits interact in complex ways, it is difficult to predict the way evolutionary changes in one trait influence changes in another trait (Leroi *et al.*, 1994).

A third limitation for both approaches concerns the effects of the genetic background of species on evolutionary changes. If two species are exposed to the same environment, they will not necessarily evolve in the same way because of differences in their genetic backgrounds. Consider the example of an insect encountering a toxic chemical in its environment which kills the insect by attacking its nervous system. Many insecticides act in this way. There are a number of ways the effects of this chemical can be overcome by evolutionary changes. First, the insect can evolve ways of detoxifying it by metabolizing the chemical to less harmful substances. Second, the insect can evolve ways of keeping the chemical out by changing the composition of its cuticle to make it less permeable to the chemical. Third, the insect's nervous system can evolve to make it less susceptible to the chemical. Finally, the insect can evolve ways of detecting the chemical in the environment and thereby develop ways of avoiding places where the chemical occurs at high concentrations. These diverse evolutionary responses have all been shown to occur when insects evolve in response to insecticides. The evolutionary changes that actually occur will depend on genes present in populations. For instance, if genes that decrease cuticular penetration of the chemical are already present at a low frequency, this trait is likely to evolve as the genes are selected and increase in frequency. An even greater diversity of responses is likely when organisms are faced with complex environmental changes. As a consequence, a trait may still be important in adaptation even if there is little association between environmental conditions and variation in the trait. A weak association may reflect the fact that changes in several traits represent alternative responses to the same environmental changes. The importance of genetic backgrounds also raises the question of how often a particular phenotype is optimal instead of being one of several alternatives.

Some criticisms have been aimed specifically at comparative approaches based on phylogenetic analyses. Many phylogenetic inferences depend on what is known as the 'principle of parsimony'. Under this principle, the present distribution of a trait is assumed to have arisen by the fewest number of evolutionary steps possible. Because of this, ancestral species will have similar characteristics to the species derived from them. This allows ancestral characteristics to be determined from those of extant species, enabling the use of techniques such as those described above for determining changes during independent evolutionary events. However, the parsimony principle is unlikely to apply if traits have been continually evolving while species diverged from each other (Frumhoff & Reeve, 1994). Characteristics of extant species may then be quite different to those that have existed during the evolution of these species. If environments are changing continuously, an ancestral species may have experienced conditions quite different to those experienced by derived species. Unless the ability of a trait to evolve is severely constrained, the phenotype of an ancestral species may bear little resemblance to those of the derived species.

It is presently not clear to what extent these factors limit the usefulness of comparative and optimality approaches. Some researchers have argued that interactions between traits will be common and obscure any association between traits and the environment. Any deductions about the adaptive significance of changes in traits would therefore require an understanding of how traits interact. Other researchers have argued that interactions between traits are not strong enough to limit the way traits evolve over a long time-scale. This would mean that comparisons between species and environmental variation are valid, and that traits are expected to evolve until they show optimal solutions to particular ecological problems.

Summary

Mechanisms that have evolved in response to changing environmental conditions can be identified by two approaches, involving comparisons between species and predictions about the optimal phenotypes organisms should possess when living in particular environments.

The comparative approach may involve comparisons of divergence among pairs of species or a number of species. When a number of species are involved, it becomes less likely that associations between traits and environments are spurious. Comparisons between species, genera or higher taxonomic levels should preferably consider phylogenetic associations among taxa. Species comparisons have been used to identify biochemical, morphological and physiological changes that have evolved in response to environmental changes.

Tradeoffs may impose long-term evolutionary constraints on species. These effects can be examined using comparisons of species, but there are several problems in interpreting associations between traits at this level.

There are various ways of identifying optimal phenotypes in populations and these have been used to predict life history patterns of organisms in variable environments. Fitness in variable environments is often defined in terms of the geometric mean rather than the arithmetic mean. When taking into account age structure in populations, the stochastic rate of increase of a population (a) is an appropriate measure of fitness. To predict the way an organism's life history changes in response to environmental conditions, the intrinsic rate of increase of a population (r) and the net reproductive rate of an organism (R_0) have been used. Some optimality applications to variable environments involve finding evolutionary stable strategies (ESSs).

Extreme conditions should select for an increased incidence of seed dormancy. This prediction is difficult to test because environmental factors can influence seed dormancy. As environmental variability increases, individuals that reproduce over a longer time interval are favoured, although this depends on age structure and whether good and bad environments occur randomly.

Variable environments may select for a reduction in the maximum number of progeny individuals can produce (conservative bet-hedging). An increasing frequency of poor years may favour asynchronous hatching of offspring and there is some evidence for this in birds. Increased dispersal ability is favoured in environments that vary in space, although this depends on whether good and bad conditions occur randomly and on age structure.

Optimality approaches have considered interactions among traits. Seed dormancy may be constrained by longevity while insect migration may be constrained by size and growth rate. Interactions between traits will influence the plastic changes organisms make when they encounter new conditions; for instance, organisms are expected to mature at a smaller size if the environment reduces growth rate because of the interaction between development time, reproductive output and juvenile survival.

Both the comparative and optimality approaches may be limited to some extent because interactions between traits and effects of genetic background can influence adaptive responses. Determining the phenotypes of ancestral species in phylogenetic analyses may be difficult if traits change rapidly in response to environmental conditions. Such problems highlight the need for complementary genetic studies within species to understand evolutionary processes.

CHAPTER 6

Extinction, diversification and evolutionary rates

Large scale environmental changes have taken place in geological time. In this chapter, we consider the effects of these changes on evolution as monitored from the fossil record. We start by considering their impact on the extinction of major species groups. We then examine their effects on the patterns and rates of evolution. Do evolutionary events occur sporadically in bursts associated with environmental changes, or are they independent of such changes? Does environmental stability influence the rate of evolution? What mechanisms can explain associations between patterns of evolution and environmental changes?

We also attempt to gain insight into how species have reacted to environmental changes. Do they evolve to counter the effects of a change, or do they simply alter their distributions or go extinct if suitable habitat is no longer available? Can the fossil record be used to make useful predictions about likely future responses of species to environmental changes?

Finally, we compare evolutionary responses in the fossil record with those in extant species. Are the genes responsible for evolutionary differences within and between populations also responsible for differences between species? If not, how are they different? These types of considerations raise the question of whether evolutionary changes within and between populations are based on the same mechanisms as those that comprise evolutionary changes in the fossil record.

Environmental change and extinctions

Extinctions occur when species are subjected to stresses and when they are unable to adapt to conditions arising from these stresses. Extinctions are an important component of the record of life on Earth since the species living today represent a very small proportion of those that have ever existed. The converse of extinction is diversification, involving the evolution of new species and higher taxonomic groups.

Extinction and diversification occur at varying rates in the fossil record throughout the Phanerozoic, that part of geological time when fossils indicate that life was abundant. There have been periods of high extinction levels,

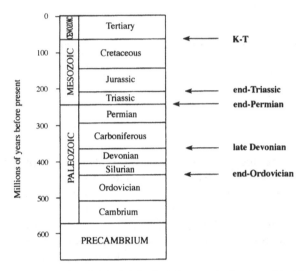

Figure 6.1. Geological time-scale showing the principal extinction events of the Phanerozoic. The arrow lengths are roughly proportional to the extinction intensity, and the five major extinction events are identified. (Simplified from Sepkoski, 1986).

followed by periods when diversification occurred at a high rate. The most extreme periods of extinction are referred to as 'mass extinction' events. The five events named in Figure 6.1 all involved the elimination of a very large number of species on a global level. Arrow lengths in Figure 6.1 are roughly proportional to the intensity of extinction, as measured by the amount of biological diversity that was lost. The most extreme of the extinctions occurred at the end of the Permian Period. The most recent and most studied extinction event is the 'K–T' event at the boundary between the Cretaceous and Tertiary periods.

It is difficult to estimate the number of species disappearing during these extinction events. This is partly because the fossil record is incomplete, and partly because individuals from different species that are morphologically similar are almost always classified as belonging to the same species in fossil studies. In the Permian event, it has been argued that around 96% of all living species could have been eliminated, while estimates for the K–T event are in the range 50 – 70% of species eliminated. These estimates are contentious and may not be particularly reliable. For instance, Briggs (1991) has suggested that the extinction rate at the K–T boundary may only have been in the order of a few per cent. He pointed out that most living species are terrestrial arthropods, vascular plants and nematodes, and that there is almost no information on extinction in these groups. Although a high proportion of species became extinct in some areas (coasts, continental shelf areas, reefs), extinction rates

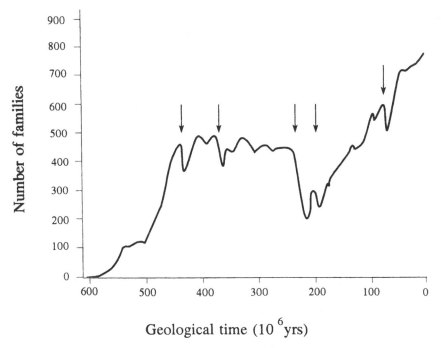

Geological time (10^6 yrs)

Figure 6.2. Changes in the diversity of skeletonized marine animals (as measured by the number of taxonomic families) over geological time. Arrows indicate the five major extinction events (From Erwin, Valentine & Sepkoski, 1987).

were low in other areas (deep seas, high latitudes). Nevertheless, there is no doubt that many groups, including entire genera, disappeared from some environments during the K–T event.

As evident in Figure 6.2, mass extinctions are usually followed by a substantial rebound in species diversity, indicating that a period of diversification tends to follow a mass extinction. This rebound is rapid on a geological timescale but may nevertheless take many years. For instance, the re-establishment of reef communities can take many million years after a mass extinction (Raup, 1991). As well as these bursts of extinction and diversification, we can see a trend towards increased diversification over time from the beginning of the Phanerozoic to the present. Thus total evolutionary diversity has increased towards the present, interspersed with occasional periods when species diversity shows a rapid decline.

Because mass extinctions have occured at the same time in diverse taxa, it is believed that they are primarily associated with global changes in physical features of the environment. These changes may be responsible for the extinctions themselves, or they may have caused secondary changes

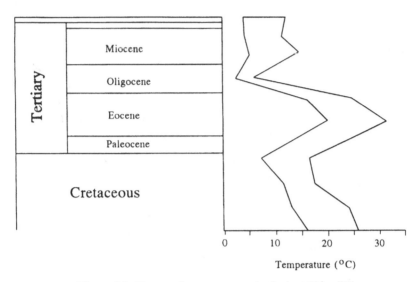

Figure 6.3. Changes in temperature in the last 100 million years determined from the ratio of oxygen isotopes in fossil shells of marine molluscs. The width of the band indicates variation in different estimates. (From Raup, 1991).

involving biological factors such as competition or predation. There is much evidence that climate and other physical features of the environment have undergone drastic changes in geological time. Chronologies of climatic change are difficult to reconstruct, but temperature data can be obtained from the distributions of fossils known to be climatically restricted, as well as from analyses of the ratios of different isotopes. For instance, Figure 6.3 gives the temperature record for the last 100 million years inferred from the oxygen isotope ratios found in fossil shells of marine molluscs. The trend is towards cooling as the present is approached, superimposed on many shorter-term fluctuations. Records from other fossils are consistent with this cooling trend.

Mass extinctions have often been associated with global climate changes. Temperature changes appear to be particularly important in both marine and terrestrial environments. The effects of cooling on mass extinctions are well-documented in the tropics. Organisms that are normally restricted to temperate regions have often expanded into tropical regions following mass extinction episodes, whereas tropical organisms have often become extinct. Recent evidence suggests that even short abrupt changes in temperature can be associated with extinction events. For instance, foraminifera (small, shelled protozoa) living on the sea floor became extinct at the boundary of the Palaeocene and Eocene, around 55 million years ago. This coincides with a sharp increase in temperature over a few thousand years (see Koch, Zachos &

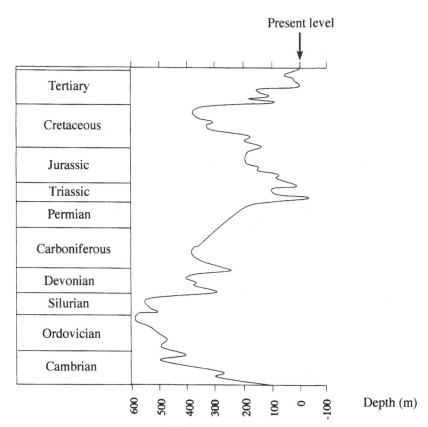

Figure 6.4. Changes in sea level during the Phanerozoic, excluding some larger fluctuations in the Pleistocene and during other glaciations. (From Hallam, 1989).

Gingerich, 1992). The exact reason for global climatic changes is not known. In some cases, it has been suggested that extraterrestrial bodies hitting the Earth may have rapidly altered climatic conditions, but it is not clear how rapid changes can be responsible for extinctions that may have happened over several million years.

Mass extinctions have also been related to decreases in sea level. Figure 6.4 provides a possible reconstruction of sea levels in the Phanerozoic. Levels have fluctuated considerably, and sea level is at present lower than at most times in the past. Sea level falls have been associated with periods of extinction, particularly for marine organisms living on the shelf area of continents. Two explanations have been proposed for this. Firstly, a fall in sea level can simply cause the disappearance of marine habitat, particularly on continental shelves. Secondly, sea level falls may cause the spread of poorly aerated water from the bottom of oceans. This influx may also have caused the extinction of marine organisms as conditions became unsuitable (Hallam, 1992).

However, falls in sea level are less likely to account for mass extinction of organisms other than those from shelf habitats. As noted above, the same mass extinction events have affected organisms from terrestrial habitats as well as from other marine habitats, so falls in sea levels are probably not responsible for extinctions of species from all areas. It is also difficult to separate the effects of changes in sea level from those involving temperature. Sea levels decrease as the amount of ice builds up due to cooling. Conversely, sea levels can influence climate because large masses of shallow water will provide some buffering against climatic changes.

A number of other factors have been proposed as extinction agents. These include changes in ocean salinity and in oxygen levels, especially in shallow marine waters (Donovan, 1989). There is also the possibility that interactions among stresses have caused mass extinctions. For example, during the K–T event, combinations of stresses are likely to have occurred, including acid rains and fluctuations in temperature, humidity and CO_2 (Briggs, 1991). Other postulated stresses during this event include darkness, giant forest fires, and atmospheric pollution resulting in acidic rain, especially because of nitric acid. A consequence of such stresses would have been a reduction in the amount of photosynthesis carried out by plants, leading to a reduction in food and widespread starvation. Collectively, all of these factors may have acted in combination with climatic changes to trigger extinctions.

Selective nature of extinctions

As we have already mentioned, groups of organisms can differ in their likelihood of becoming extinct. In the case of extinctions associated with a temperature decrease, tropical biota are more likely to become extinct than temperate biota. When cooling occurs, refuges are unlikely to be available to warm-adapted organisms as their distributions contract towards the equator. In contrast, when warming occurs, refuges are more likely to be available for cold-adapted organisms because cold conditions persist at high latitudes and at high elevations. This is consistent with relatively high rates of extinction recorded in some tropical habitats at low latitudes, particularly in tropical reef communities.

The probability of a group of organisms becoming extinct may also be related to its geographic distribution. From an analysis of the geographical ranges of molluscs from the Cretaceous, Jablonski (1987) found that species achieved their ranges relatively early in their geological histories. Because of this, ranges can be regarded as a property of species rather than being continually altered. Jablonski (1987) also found that when pairs of related species

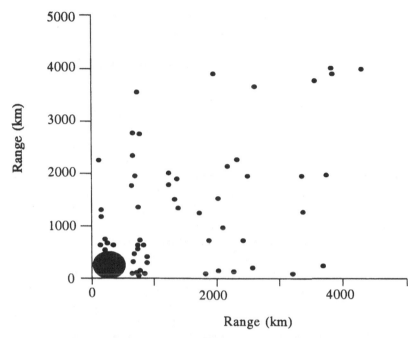

Figure 6.5. Association between the geographic ranges of pairs of
closely-related gastropod species from the Late Cretaceous. Each point
represents a pair of species, while the large circle represents more than
30 species pairs (Simplified from Jablonski, 1987).

were considered, their geographic ranges were correlated to some extent
(Figure 6.5). This means that when one species had a wide geographic range,
its relative was also likely to have a wide range. This suggests that geographical ranges of species are not randomly determined, but are constrained by
their evolutionary lineage.

Jablonski and others have found that the length of time that mollusc species
persist in the fossil record are positively correlated with their geographical
range. Widespread species appear less prone to extinction than restricted
species. However, this relationship was not evident at the time of the K–T
event when a wide distribution did not protect species from extinction
(Jablonski, 1991). Perhaps environmental stresses causing the mass extinctions
were so widespread during this event that there were no safe refuges.
Nevertheless, widespread genera of molluscs survived the K–T event better
than those with restricted distributions. This indicates that wide geographic
distributions gave some protection against extinction at a higher taxonomic
level.

We can postulate several reasons why widespread genera and species are less
prone to extinction than those with restricted distributions. First, refuges may

be more readily available to widespread groups because refuges are more likely to exist over a large geographic area than a restricted area. Second, widespread species may have higher innate levels of resistance to environmental stresses because they normally encounter a wider range of conditions. This is evident, for example, in the higher levels of stress resistance exhibited by widespread species of extant groups of insects such as *Drosophila*. In this genus, cosmopolitan species have high levels of resistance compared to relatives with restricted distributions. Third, a large geographic range may help protect against stochastic changes in population size due to local catastrophes. Global environmental changes associated with extinction events may induce local catastrophes such as fires and storms that wipe out populations. When organisms are distributed over a large area, they are less likely to become extinct as a consequence of such catastrophes because not all populations are exposed to them. Finally, widespread species will often be subdivided into populations adapted to a wide range of local conditions, whereas restricted species will be adapted only to the local conditions they normally encounter. This means that genes important in adaptation to a range of stresses are more likely to be present in widespread species, increasing their potential for adapting to changing conditions.

Apart from being associated with geographical distributions, the likelihood of species becoming extinct is also associated with body size. If we consider terrestrial vertebrates, there is evidence that large organisms tend to be more extinction prone than small ones. Well-known examples are the disappearance of the dinosaurs in the K–T event and the extinction of large mammals in the Pleistocene. Body size may also be important in other groups. Large invertebrates are often more prone to extinction than small invertebrates, as in the case of mollusc taxa (Hallam, 1990). Body size has been related to recently-documented extinctions of bird populations from islands around Britain (Pimm, Jones & Diamond, 1988). There are several reasons why large animals may be more prone to extinction. These include the tendency of large species to have small population sizes and low inherent rates of increase in size, making them more prone to environmental catastrophes.

Some groups of organisms may be relatively less prone to extinction events because of the characteristics they possess rather than because of their distribution or size. For instance, terrestrial plants appear to be less affected by extinction events than other groups of organisms. Plants are more likely to survive extinction events because of characteristics such as vegetative reproduction, indeterminate growth, potential long dormancy of propagules, and efficient dispersal to facilitate migration (Traverse, 1988). As a consequence, terrestrial plants may evade stress by entering dormant stages, thereby evading starvation at unfavourable times.

Insects are another group of organisms that may have low extinction rates. For instance, insects emerged relatively unscathed during the extinction event at the end of the Cretaceous. Insects often have long-term associations with plants, and also possess mechanisms for reducing effects of environmental stresses, including entering dormant stages of a life cycle, and evading stressful habitats by switching resources or migrating.

Dormant life cycle stages may also be important in plankton. One feature of the K–T event is the selective extinction of marine plankton groups. In groups known as the coccolithophorids, radiolaria, and foraminifera, extinctions of genera exceeded 70, 80 and 90% respectively. In contrast, planktonic diatoms only suffered an extinction rate of around 20–25%. Diatoms probably survived relatively unscathed because they can evade periods of nutritional depletion by forming a resting spore. These provide a way of evading stressful conditions that can be lethal for actively growing diatoms (Kitchell, Clark & Gombos, 1986).

In the above cases, low rates of extinction appear to be associated with an ability to evade stressful conditions. In addition, the risk of extinction may be reduced by an ability to survive these conditions. Rhodes & Thayer (1991) have postulated that global climatic changes at times of mass extinctions were likely to generate periods of reduced primary productivity. Species adapted to these conditions should therefore be less prone to extinction. This hypothesis was tested by examining bivalves from the K–T boundary. They found that suspension-feeding bivalves which live on phytoplankton were more prone to extinction than bivalves consuming detritus and bacteria. This was expected because a reduction in primary productivity should decrease the availability of phytoplankton more drastically than that of detritus.

Stress resistance was also favoured in other organisms (Rhodes & Thayer, 1991). Among suspension feeders, extinction was more marked in swimming and burrowing bivalves than in sedentary animals. In addition, shelled bivalves suffered greater extinction rates than brachiopods, another group of shelled organisms. These findings are in agreement with the reduced productivity hypothesis. Sedentary animals require less energy than active animals, enabling survival under low levels of productivity. Brachiopods are more resistant to starvation than shelled bivalves because they have lower metabolic rates. Extinctions were therefore concentrated among groups of animals with starvation-susceptible feeding modes, active locomotion, and high metabolic rates.

To summarize, it appears that mass extinctions are associated, at least indirectly, with drastic environmental changes. These changes seem to favour the survival of widespread taxa with smaller body sizes. Organisms resistant to stresses appear to be relatively less prone to extinction. This resistance may involve low resource requirements or evasion of stresses via resistant life-cycle stages. These mechanisms were probably selected to counter seasonal periods of unfavourable climatic conditions, or to enable survival in environments that are constantly stressful. Because of this, the concept of tradeoffs between environments discussed in Chapter 4 may apply to long-term evolutionary persistance as well as to short-term environmental changes. Stress-resistant genotypes that are at a disadvantage in favourable environments may be selected at extremely stressful times associated with mass extinction. We should emphasize that dormant stages are not themselves adaptations to extremely rare extinction events, but organisms having these stages are fortuitously more likely to survive these events!

Mass extinctions: creative or destructive?

Since Darwin, evolutionary change has often been viewed as the replacement of one lineage by another via competition. Macroevolution was seen as continually increasing the competitive ability of organisms as lineages with 'superior' competitive abilities replace others. This view of evolution predicts that changes in the fossil record will appear as in Figure 6.6(*a*), where one lineage gradually replaces another over a period of a few million years. As one lineage expands, the other contracts because it is outcompeted.

However, this view is now being challenged because the type of fossil pattern associated with competitive replacement seems to be rare (Benton, 1987). More commonly, paleontologists have found that the expansion of a lineage does not coincide with the gradual contraction of another (Figure 6.6(*b*)). Many cases of evolutionary change, that were previously viewed as the gradual replacement of one lineage by another, are now considered to represent more abrupt evolutionary changes unlikely to be driven only by competition. For instance, in the Mesozoic, brachiopods were replaced by bivalves, which include extant molluscs. This replacement was not gradual, but occurred rapidly during the mass extinction event at the end of the Permian. Bivalve diversity also decreased during and following this event, but it rapidly recovered. This pattern of evolution does not represent a gradual replacement of brachiopods as predicted by competition, but an abrupt replacement triggered by a global environmental change.

This has led to suggestions that mass extinction events may have a role in 'stimulating' evolutionary change. As we have already mentioned, periods of diversification normally follow mass extinctions. This can be seen in Figure 6.2 which plots the number of families of skeletonized marine invertebrates through geological time. There was an initial diversification in the Cambrian around 550 million years ago when the remains of complex and diverse organisms suddenly appeared in the rock record. This is shown escalating rapidly into the Ordovician Period (around 500 million years ago) until the end of this period when the first mass extinction took place. Thereafter diversification rapidly increased to be curtailed by the second mass extinction, and so on. Mass extinctions therefore appear to be associated with periods of evolutionary change.

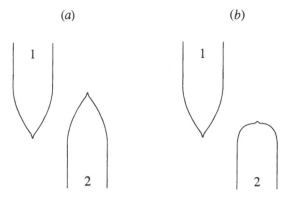

Figure 6.6. Changes expected in the fossil record when species replacement involves competitive interactions (*a*) or when species replacement is independent of competition (*b*). (Simplified from Benton, 1987).

How can extinction events stimulate evolutionary change? One view is that extinction events remove poorly-adapted species, and that this allows for the expansion of well-adapted species. An example often cited in support of this view is the diversification of mammals which followed the extinction of dinosaurs in the K–T event. Because mammals are considered to be more complex than the large reptiles, this replacement is often viewed as an example of evolutionary progress. However, mammals and dinosaurs coexisted for more than 100 million years before the K–T event, and it is not clear why mammals did not replace dinosaurs over this time if they were relatively better-adapted.

An opposing view is that mass extinctions can eliminate many highly-adapted species. As we discussed above, extinctions are caused by severe stresses that are extremely rare. Extinction events often eliminate specialist species with restricted distributions that are adapted to specific conditions. After extinction events, these may be replaced by widespread species that tend to be more archaic, less morphologically complex, and have the ability to resist or evade stresses. Extinction events will therefore eliminate, at least initially, many well-adapted species. Any 'creative' role for mass extinctions must involve their indirect effects rather than any direct effects.

Several ways in which mass extinctions may indirectly stimulate evolution have been proposed (Table 6.1). We first consider the effects of extinctions on predation. Vermeij (1987) has suggested that extinctions may facilitate evolutionary changes in the morphology of organisms, because extinctions influence levels of predation and other biotic interactions. This is based on Vermeij's view that changes in the fossil record in the morphology of gastropods, cephalopods and other groups are largely due to predation. For

Table 6.1. *Possible ways in which mass extinctions indirectly increase rates of evolutionary divergence*

1. Mass extinctions decrease predation pressures, enabling new evolutionary novelties to become established; subsequent radiations driven by renewed predation pressure

2. Mass extinctions decrease competitive interactions, allowing previously non-competitive species to radiate

3. Ecological space cleared following mass extinctions is occupied by novel forms adapting directly to environmental conditions

4. Stressful conditions associated with extinction events result in the expression of increased genetic variability

5. Stressful conditions change the nature of the fitness landscape, resulting in peak shifts

instance, gastropods have a wide range of features that help to resist predatory attacks. These include large, thick shells and barriers such as teeth to cover apertures. Such defence mechanisms have evolved and become more elaborate over time, as predators have become more efficient at overcoming defences. Morphological changes in gastropod evolution may therefore represent, at least partially, an 'escalation' process between gastropods and their enemies.

Vermeij (1987) suggested that extinctions have a creative role because they cause a cessation of this escalation process as many enemies are eliminated. Predators and well-protected prey may be particularly susceptible to extinction. For instance, well-armoured molluscs appear to be more susceptible to extinction than less protected species, perhaps because the energetic costs of producing armour place these organisms at a disadvantage under stressful conditions when energy needs to be expended to ensure survival (see Chapter 4). The elimination of escalation means that organisms no longer have to be as well protected against predators. Vermeij envisaged this process as freeing up lineages to follow novel evolutionary pathways. As enemies become re-established, the escalation process recommences, and new pathways allow different and more complex sets of defences to be developed. Extinctions driven by environmental factors are therefore seen to have an indirect role in triggering a further cycle of escalation.

This process may not be restricted to gastropod groups. For instance, morphologically complex species of oceanic plankton are more likely to become extinct than simple forms during extinction phases (Lipps, 1986). Complex forms are probably better protected against predation, but may be at a disadvantage during periods of decreased resources.

A second hypothesis is that evolutionary changes are triggered by the effects of extinction events on competition. As well as removing species from the pressures of predation, extinction events may allow the replacement of one species by another via a decrease in competitive interactions. This hypothesis was favoured by Hallam (1990) in explaining evolution in bivalves and ammonites in the Jurassic and Triassic periods, a time when two mass extinction events occurred (Figure 6.1). Species of these molluscs are distinct, having maintained the same morphology for many millions of years with the exception of gradual changes in size. These periods of relatively minor changes are known as periods of 'stasis'. They are interspersed with periods of rapid change, referred to as 'punctuations', when an ancestral species is replaced by another related species which is either a descendent or an immigrant.

Hallam (1990) agued that one mollusc species normally excludes another by competition, accounting for periods of stasis. Occasionally, after mass extinction events have occurred, habitats are substantially vacated by one species. This allows a related species to take over, leading to an abrupt change in the fossil record. We should emphasize that such changes are not a consequence of one species displacing another by competition, because two species would then be expected to coexist for some time (Figure 6.6). Hallam (1990) therefore used the term 'pre-emptive' competition to describe this phenomenon, as opposed to 'displacive' competition. Under displacive competition, a species arriving later would be able to displace another species already occupying an ecological niche. However, under pre-emptive competition, a species occupying an area is able to competitively exclude other species.

In both the predation and pre-emptive competition hypotheses, mass extinctions have the role in generating ecological opportunities for species. Distributions of species are normally seen as being restricted by biotic interactions, but drastic reductions in distributions provide ecological niches that can be filled by new forms. As such, radiations following extinctions are not seen as being a direct consequence of selection for new taxa, but extinctions act as an indirect stimulant of evolutionary change.

We can see the effects of vacant space on evolution in oceanic islands. Organisms reaching islands often undergo dramatic evolutionary radiations into unoccupied habitats. The absence of competition on islands is considered a prerequisite for these radiations. This enables lineages to adapt to new habitats occupied elsewhere by other species. For instance, one of the best-documented examples of evolutionary changes on islands is the radiation of finches in the Galápagos. The fourteen finch species in these islands have a range of beak sizes and shapes, and this is associated with differences in diet.

Species with long and pointed bills probe flowers, foliage or woody tissues, and feed on nectar or small arthropods. Other species with bills deep at the base can crush hard foods such as seeds. These differences are likely to have evolved after an ancestral finch species reached the Galápagos, and adaptive changes occurred in response to vacant habitats due to the absence of other species (Grant, 1986).

Although vacant ecological space seems to be important in generating evolutionary changes, it is still not clear if competition and predation play a large role in these changes (Masters & Rayner, 1993). If the absence of competition or predation pressures is needed for evolutionary novelties to become common, why should the presence of the same pressures stimulate further evolutionary change? After a radiation into vacant ecological space, biotic interactions will become increasingly intense as the space fills up. As this happens, biotic interactions will probably cause the extinction of some novelties and expansion of others. However, if the renewed biotic interactions themselves are to result in further evolutionary changes, they need to drive successive replacements of one form by another. Why should such a process occur when Vermeij and others have argued that biotic interactions normally promote stasis?

It is possible that biotic interactions can promote evolutionary changes when organisms are expanding to occupy vacant space. Rosenzweig & McCord (1991) have argued this view in relation to competition, and propose the following scenario. Consider a situation where a species (say species A) occupying a habitat becomes extinct, whereas another species (species B) survives. Any individuals from the surviving species that resemble species A would be at an advantage, because they can occupy the vacant niche and do not have to compete with other members of species B. By moving into a vacant space, new forms of species B can increase their fitness by avoiding competition with other members of the same species. Competition within species could therefore facilitate the evolution of new forms in vacant ecological space. Eventually, new forms may become reproductively isolated from species B, a point which we return to later.

Although this scenario suggests a possible role for biotic interactions, we still need to separate their effects from evolutionary changes directly involved in adapting to new environments. A new form of species B may have a higher fitness in vacant space irrespective of competition. For instance, a bird that can crack large seeds or that can tolerate drier conditions may have a higher fitness than other individuals because it has access to a wider range of resources, and not because it avoids intense competition.

The above suggests that environmental changes leading to mass extinctions may have an important role in clearing ecological space, but it is still not clear why new evolutionary forms should emerge at these times. If competition and predation pressures are reduced, then new evolutionary lineages with a poor competitive ability, or low predation resistance could persist for longer, but why shouldn't these be eradicated eventually? We will return to this question after examining the effects of different environments on patterns of evolutionary change, because these provide additional insights into factors promoting evolutionary radiations.

Morphological variability and rugged fitness landscapes

There are two other reasons why extinction events trigger evolutionary change. In light of the discussion in Chapter 2, one possibility is that by inducing variability the stressful periods directly increase the rate of evolution of morphological traits. Many traits are invariant because they are locked into developmental pathways that are buffered against environmental variability. Because periods of intense and unusual stress disrupt development and dramatically increase the expression of variability in such traits, conditions during mass extinctions could have a direct role in promoting the emergence of evolutionary novelties. There would undoubtedly be many occasions during these events when organisms are stressed sufficiently for developmental processes to break down, resulting in novel forms. Although most novelties would have a low fitness especially under stressful conditions, some may persist because they can use a previously vacant niche.

The second possibility is that environmental changes influence the fitness landscape associated with traits that form parts of interacting networks. Kauffman (1993) has recently argued that stasis in such traits arises because of the way genotypes and phenotypes influence fitness. He starts with a model where evolutionary transitions between alternative genotypes and/or phenotypes involve a series of interlinked steps. A simple example is given in Figure 6.7(a). Assume that A_1 and A_2 are alternative alleles or phenotypes, and similarly for B_1/B_2, C_1/C_2 and D_1/D_2. For an organism to move from one state $(A_1B_1C_1D_1)$ having a low fitness to another $(A_2B_2C_2D_2)$ having a relatively higher fitness requires a series of steps. Now if these states differ in fitness, a population can become stuck in a particular state when making the transition. For instance, assume that fitness values of the different states correspond to those in Figure 6.7(b). In this case, the fitness of an allele depends on which allele is present at the other loci, reflecting epistatic interactions between the alleles. A population reaching the $A_1B_1C_2D_2$ state may find it difficult to move further because a mutation resulting in A_1 changing to A_2 or B_1 changing to B_2 will result in a genotype/ phenotype with a relatively low fitness. A population may therefore remain in a particular state and fail to reach the highest fitness state.

These types of interactions between genes and phenotypes can produce a

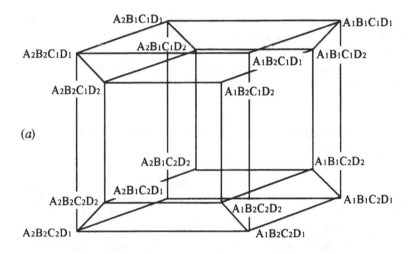

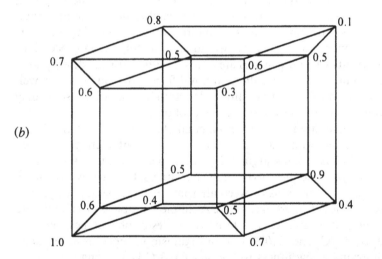

Figure 6.7. Possible combinations of alternative alleles and their fitness at four loci in a haploid organism. In (*a*), genotypes are presented in a cubic space and joined by lines separating single mutation events. In (*b*), fitness values have been assigned to the genotypes. Evolution from the genotype with lowest fitness ($A_1B_1C_1D_1$) to the genotype with highest fitness ($A_2B_2C_2D_2$) may be difficult because of intermediate steps with a relatively high fitness. (Modified from Kauffman, 1993).

fitness 'landscape', which is a fitness surface where there are peaks representing states with a high fitness, surrounded by troughs representing states with a low fitness. Sewall Wright (1931) proposed the existence of such landscapes many years ago. Complex landscapes rely on complex interactions between genes or phenotypes in producing fitness. In Kauffman's (1993) model, the complexity of landscapes is determined by two factors, the number of components of a system (N) and the number of interactions between these different components (K). For instance, the epistatic interaction we outlined in Table 2.2 involving two loci would have the values $K=1$ and $N=2$. As K increases and as long as N is large, fitness landscapes can become extremely complex. The presence of numerous peaks makes it difficult for populations to reach the peak with the highest fitness. Conversely, when $K=0$, there is a single fitness peak due to the absence of interactions between components. If the A, B, C and D components in Figure 6.7 only contribute in an additive manner without epistasis, there would be a single peak of high fitness ($A_2B_2C_2D_2$) and the population would move rapidly to this peak from its original state.

Evolutionary rates will therefore depend on the types of landscapes encountered by organisms. It will be difficult for a population to move from one peak to another when peaks are steep and landscapes are rugged because fitness differences between alternative forms are large. High peaks in complex landscapes lead to stasis. Any mutations that occur in such landscapes are only likely to be favoured if they cause small phenotypic changes. Conversely, the absence of any peaks will also lead to a slow rate of evolution. A smooth fitness surface implies small fitness differences between genotypes/phenotypes and relatively weak natural selection.

If a population is in a rugged landscape with high multiple peaks, further evolution may not occur unless the fitness surface itself changes. This can happen when environmental conditions are altered, producing a new set of peaks. Major mutations can then cause populations to shift rapidly to new peaks, and minor mutations can help populations to move up fitness gradients once they reach these peaks. Evolution in complex systems is therefore expected to be constrained under the NK model unless the fitness surface is altered. Drastic environmental changes trigger evolutionary changes by facilitating peak shifts.

These arguments may help to explain patterns of radiations and stasis in the fossil record associated with extinction events. Kauffman (1993) argued that, at the time of the Cambrian explosion when the major animal body plans emerged, multicellular organisms were poorly adapted to their environments. This means that fitness peaks were not particularly high. As a consequence, mutants leading to different body plans were more likely to survive as they moved to different regions of the fitness surface, resulting in peak shifts. However, in subsequent extinction events, multicellular organisms were relatively better adapted to their environment, resulting in higher fitness peaks and an increasingly rugged landscape. This made it harder for fundamental changes in the body plan to occur because organisms were trapped at high fitness peaks with respect to traits determining body plans. Evolutionary changes were therefore confined largely to shifts within peaks for such traits.

Major environmental changes may therefore play a role in evolution by both triggering morphological variability and by altering fitness surfaces so that peak shifts can occur on a rugged fitness landscape. These hypotheses do not require vacant ecological space, but instead rely on the effects of stress on epistatic interactions and how these interactions relate to fitness. Unfortunately, while complex interactions among loci and phenotypic components undoubtedly occur in the development of organisms, it is difficult to test these hypotheses directly from the fossil record.

Patterns of evolutionary change in different environments

There is evidence that new evolutionary forms are more likely to arise in some types of environments than others. For instance, Jablonski & Bottjer (1990) considered invertebrate fossils along marine gradients extending from the shoreline. They divided habitats into five regions: nearshore, regions encompassing the inner, middle, and outer shelf, and regions on the slope of the shelf and in the deep basin. The nearshore and inner shelf categories (classified collectively as onshore) are more frequently disturbed than the remaining habitats (classified collectively as offshore). Higher taxa of invertebrates in post-Paleozoic time arose much more frequently in the onshore habitats than in the other areas. This is especially true for invertebrate groups living on the sea floor with a high potential to be preserved as fossils. The new lineages then expanded offshore where communities tended to be more archaic than onshore.

Primary producers are abundant in onshore environments, and energy is continually introduced into the environment via photosynthesis to provide new resources that can be exploited further down the food chain. In contrast, food tends to be less abundant in offshore habitats because both organic levels and oxygen content decrease towards the deep ocean floor. In deep seas where primary producers are absent, the availability of energy is lower because it depends on the rate at which nutrients are introduced from other environments with primary production. New taxa therefore tended to originate in disturbed environments where selection pressures continuously change, and where food is not limiting (Jablonski & Bottjer, 1990).

We can find evidence for analogous evolutionary patterns in terrestrial plants. Archaic plant taxa are found throughout the Phanerozoic in swamps and other wetlands that are wet and acidic, and that lack oxygen and nutrients. Evolutionary innovations are characteristic of drier areas and habitats that have been disturbed and periodically exposed to moisture stress. For instance, DiMichele & Aronson (1992) considered flora from the Late

Carboniferous and Early Permian periods (around 280 million years ago). They found that orders of early vascular plants originated in drier habitats that are variable and occasionally stressed. These orders then became prominent during the Late Permian and Mesozoic (more than 40 million years later). Flowering plants also originated in disturbed habitats and were originally weedy and fugitive species.

These fossil patterns can be related to the current association between novel plant forms and habitat type. The flora of southwestern Australia provides one example. This region has a particularly high concentration of plant species, many of which do not occur elsewhere in the Australian continent. Hopper (1979) identified three major habitat categories: (1) a high rainfall zone where conditions are humid, (2) a transitional rainfall zone characterized by erosion and where climatic stresses occur regularly over a longer time-scale of several years, and (3) a low rainfall zone characterized by continuously stressful conditions. Up to three times as many species occur per given area in the transitional zone compared to other zones. Populations in the transitional zone have been fragmented and have fluctuated in size as a consequence of erosion, fire and climatic effects on moisture conditions. These unstable conditions that are sporadically stressful appear to have favoured the evolution of diverse plants. In contrast, the high rainfall zone has been relatively stable over millions of years. There are many relict species in this area, suggesting a relatively slow rate of evolutionary change.

Another region that fits this pattern is the Succulent Karoo Region, an arid area in southern Africa (Ihlenfeldt, 1994). The environment is characterised by a variety of soils and sharp changes in local climatic conditions and available soil moisture. Conditions have fluctuated over time due to long-term climatic cycles. This unstable area is associated with a high degree of evolutionary divergence in a particular group of succulent plants known as the Mesembryanthemaceae. Rapid evolutionary divergence must have occurred in this group because the Karoo region has only developed in the last 5 million years. Despite generally arid conditions, the number of plant species per unit area in the region is high and approaches that of rainforests.

There is evidence for a slow rate of morphological evolution in environments that are continuously stressed. In evolutionary lineages representing 'living fossils', observable change is arrested over vast time periods that can encompass mass extinction events. Organisms in such lineages often live in harsh environments. For instance, horseshoe crabs, that have existed virtually unchanged in morphology for 200 million years, function in water varying widely in salinity. These organisms can endure large swings in temperature and oxygen levels, and their hardiness is confirmed by their high level of resistance to chemical pollutants. Similarly, the brachiopod, *Lingula*, and oysters and mussels that have survived from the Palaeozoic era tend to be found in stressful environments. Ward (1992) emphasized that the stress levels experienced

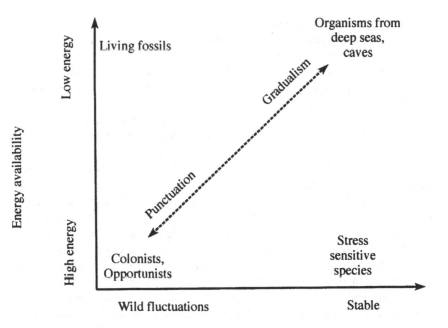

Increasing environmental stability

Figure 6.8. The interaction between the size of long-term
environmental fluctuations over thousands of years and the amount of
metabolic energy available for organisms above maintenance and
survival. The four corners represent extremes. The diagonal line
indicates the patterns of evolution associated with the extremes ranging
from phyletic gradualism to punctuational change. (Modified from
Parsons, 1993).

by these organisms are so high that their predators and competitors cannot
survive. Living fossils in harsh environments are therefore not exposed to
strong biotic interactions.

It may be possible to make some generalizations about the above patterns.
The potential for evolutionary change can be expressed by two interacting
continuums: (1) the magnitude of environmental fluctuations, and (2) the
availability of energy in an environment as determined by its productivity. We
should emphasize that environmental fluctuations in this diagram are
expressed over periods of hundreds or thousands of years rather than within
a generation or a few generations. When energy availability and the degree of
environmental fluctuation are combined, there are four extremes (Figure 6.8)
with a different potential for the generation of new evolutionary lineages.

(i) Widely fluctuating environments, low energy availability

This includes environments where many 'living fossils' are found. The environments are continually stressful and fluctuate repeatedly. Low energy environments may occur at high latitudes where harsh winter conditions are interspersed with short growing seasons. They may also occur at the upper reaches of rocky tidal zones where organisms are repeatedly exposed to desiccating conditions after the tide retreats. In these conditions, there is likely to be physiological adaptation to cope with stressful conditions, but major evolutionary changes in response to biotic factors are unlikely because low population densities and the need for organisms to expend most of their resources on survival will limit interactions among species. Organisms in this environment need to be generalists, adapted to a wide range of conditions. If these environments persist, they are likely to be associated with lineages that do not change much. If the environments change, organisms will probably alter their distributions to track suitable habitats rather than evolve.

(ii) Stable environments, low energy availability

This combination is found in environments such as deep seas and caves where there is not much primary productivity and conditions are relatively constant over thousands of years. Low metabolic rates may allow organisms to persist in these environments in the face of stresses from anoxia and low resource availability. When organisms first enter these environments, substantial evolutionary change can occur. For example, organisms undergo predictable and substantial changes when they become adapted to caves (Howarth, 1993). In cave insects, the need for energy economy is indicated by a reduction and simplification of body parts such as compound eyes, ocelli and chitinous structures, and a weak ability to fly. In marine basins, a diverse fauna of benthic invertebrates with calcareous bodies occurs higher up in basins. As the oxygen level falls, this gives way to a fauna consisting of animals that are small and have soft bodies. Despite these adaptive changes, major evolutionary novelties do not seem to develop within these habitats and species tend to be relict.

(iii) Widely fluctuating environments, high energy availability

This combination represents environments where opportunists and colonists are found. These include terrestrial environments where fire, soil movements or other disturbances change environmental conditions repeatedly, or near-shore marine environments where falls in sea level can abruptly change conditions. Populations will fluctuate in size as environments change, and will often face periodic extinction. This type of environment has been associated with the development of evolutionary novelties, as in the case of onshore habitats. There may be a spate of extinctions when severe environmental changes occur, and only generalist species with inert or resting phases survive. Biotic interactions can be important in these environments, but not on a continuous basis because of disturbance associated with environmental instability.

(iv) Stable environments, high energy availability

These environments are relatively stable over thousands or even millions of years. Organisms living in them are vulnerable to rare stresses that are extreme because organisms are not adapted to such conditions. Biotic interactions can be intense, and adaptive changes in response to predators and competition are common. These environments are less likely to be associated with major evolutionary novelties than (iii), although species diversity can be high, as in some tropical forests. This is because fine-tuning for adaptation to different resources can occur where abiotic conditions are not extreme. The efficient use of available energy can be conducive to adaptive radiations (Parsons, 1994).

Apart from influencing the rate of production of new evolutionary lineages, environmental variation may also affect patterns of evolution. There are two main models of how morphological changes occur in the fossil record. On the one hand, morphological changes may occur discontinuously as proposed by Eldredge & Gould (Eldredge & Gould, 1972; Gould & Eldredge, 1993) and others. A rapid morphological shift may take place on some occasions, resulting in an evolutionary pattern known as 'punctuated equilibrium', where periods of minor morphological change are interspersed with short periods of major change (Figure 6.9(*b*)). On the other hand, morphological change may occur in a more continual and less discrete manner within a lineage. This is known as 'phyletic gradualism' (Figure 6.9(*a*)). It should be emphasized that evolutionary changes do not always fall into these two extremes, and changes within a lineage may show a complex range of patterns encompassing gradualism and periods of punctuation.

When environmental conditions show a narrow range of fluctuations over a period of hundreds or thousands of years, and energy/resource availability is low, evolutionary changes may follow a phyletic pattern rather than a punctuated pattern. Sheldon (1992) provided evidence that trilobites from a site in Wales fit this situation. These animals thrived and evolved in a marine environment several hundred metres deep. Long-term environmental fluctuations are likely to be minimal in this environment, and the availability of energy is low because of low levels of primary productivity. This is a relaxation of the extreme situation represented by (ii) above. As expected, Sheldon (1992) found relatively gradual changes in the morphology of the trilobites as in Figure 6.9(*a*).

In contrast, major environmental changes as in (iii) above may be associated with rapid morphological shifts. Normally, extreme fluctuations in the

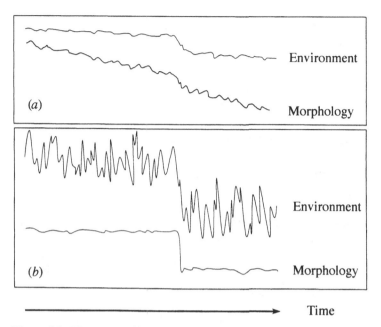

Figure 6.9. Diagrammatic representation of the hypothesis that, over geological time-scales, gradual, continuous phyletic evolution is characteristic of relatively narrowly fluctuating environments (*a*), and that stasis with rare events of punctuational change tends to prevail in more widely fluctuating environments (*b*). (Modified from Sheldon, 1992).

environment on a time-scale of hundreds or thousands of years are associated with a low level of evolutionary change, but a period of rapid change may ensue when a major shift in an environment occurs. For instance, stasis tends to prevail in marine invertebrates living on the floor of shallow waters, which is a disturbed environment. However, as we previously discussed, the first appearances of new taxa tend to occur in these onshore habitats where food is abundant (Jablonski & Bottjer, 1990). The appearance of novelties, coupled with periods of stasis, would tend to produce a punctuated pattern of evolution as in Figure 6.9(*b*) rather than gradual change.

In summary, we have described how radiations and evolutionary patterns in the fossil record may be related to environmental conditions. Evolutionary stasis appears to prevail in stressed habitats where environmental conditions fluctuate widely in geological time and primary productivity is low. When fluctuations are smaller, evolutionary patterns may approximate more closely to phyletic gradualism. Punctuated patterns are characteristic of disturbed environments where primary productivity is relatively high. It is from these environments that evolutionary novelties are likely to arise.

Table 6.2. *Explanations for associations between evolutionary patterns and the environment*

Explanations not involving speciation

1. Evolutionary patterns determined by intensity of biotic interactions which varies in different environments

2. Clearing of ecological space occurs in fluctuating environments and promotes evolution

3. Continuous environmental fluctuations in geological time prevent adaptations

4. Intermittent stresses increase genetic variability normally unexpressed because of developmental constraints

Explanations requiring speciation

5. Speciation in small, isolated populations increases likelihood of fixing major genetic changes

6. Speciation following population fragmentation facilitates fixation of geographical variation

7. Speciation into vacant ecological space increases rate of morphological evolution

Patterns of evolution: explanations

Why do patterns of evolution vary in different types of environments? Two types of explanations have been put forward. The first type is based on the possibility that the environment influences rates of evolutionary change independent of speciation. The second type assumes that the environment influences morphological evolution indirectly by affecting rates of speciation. We can group several hypotheses under these alternatives (Table 6.2), some of which are linked closely to the hypotheses about stimulatory effects of mass extinctions that we discussed previously (Table 6.1).

1. Environmental effects independent of speciation

The first hypothesis in this category is that morphological evolution depends on the intensity of biotic interactions, and that these interactions in turn depend on environmental conditions. In considering the evolutionary stasis found in stressed environments (situation (ii) above), Vermeij (1987) pointed out that food and other resources are limiting in these environments because of low levels of primary productivity. The metabolic rate of organisms needs to be low for survival, and much of the energy acquired by organisms will need to be spent on survival rather than reproduction. In addition, these environments will differ in other ways from those where food and other resources are plentiful. Low levels of resources will reduce population densities and the intensity of competition. Fewer resources will also mean fewer natural enemies. Biotic interactions will therefore be reduced. Vermeij (1987) argued that this, in turn, may decrease rates of evolutionary change in morphological characteristics such as skeletal defenses that are selected as a consequence of biotic factors. In contrast, evolutionary changes in morphological traits are expected in high energy environments where biotic interactions can be intense.

This argument is based on the assumption that much of the morphological change we see in the fossil record is driven by biotic interactions. As mentioned in the discussion of mass extinctions, this is far from clear and even unlikely. The fact that radiations occur after mass extinctions when biotic interactions are weak suggests that evolutionary change is often initiated by other factors. If biotic interactions drive morphological evolution, rates of morphological change are expected to be greatest in stable environments where productivity is high, rather than in unstable environments where novelties tend to evolve.

The second hypothesis is that evolutionary change is more likely in environments where ecological space intermittently becomes vacant. We have already presented this hypothesis with respect to mass extinctions. Supporting evidence includes the rapid evolutionary diversification that can take place when stresses are suddenly removed, permitting the occupation of new habitats that had previously been unoccupied. For instance, this sort of process has been postulated at the Precambrian – Cambrian transition (around 550 million years ago), where there was an enormous burst of evolutionary novelties. These novelties resulted in new phyla, classes, orders and families. During this transition, there were substantial changes in the chemical composition of oceans, including the release of large quantities of phosphorous into shallow waters (Conway Morris, 1987; Brasier, 1991). Phosphorus is a vital component in the normal functioning of cells, and when it occurs at low levels it limits productivity in marine environments. The removal of this limiting factor would have provided evolutionary opportunities for adapting to new environments. However, there are other possible explanations for the

Cambrian explosion, such as the effects of stress on a fitness landscape with low peaks mentioned earlier.

The availability of vacant ecological space may help to account for the development of evolutionary novelties in some environments. As we mentioned above, novelties tended to arise in environments that periodically experienced environmental disturbance. The environmental changes would have been large enough to cause local extinctions or migration of susceptible species, resulting in vacant space once conditions ameliorated. Following our earlier discussion, this vacant space may have led to evolutionary changes as organisms adapted to new conditions or evaded the effects of intraspecific competition.

There is good evidence that rapid adaptive radiations can occur as vacant space is colonized following an environmental change. In the northern Andes in tropical Venezuela, Colombia and Equador, recent glacial periods have resulted in a new set of environmental conditions associated with high humidity, low temperature and frequent frost. This has led to rapid evolutionary changes in some plant groups that have colonized this region. For instance, in a plant genus known as *Espeletia*, a recent radiation has resulted in 130 species within the region, varying from branched trees to species that form giant rosettes capable of surviving under low nutrient conditions (Monasterio & Sarmiento, 1991).

The third hypothesis in Table 6.2 may help to explain the evolutionary stasis that is often observed in lineages from environments that continually fluctuate in geological time. If environmental changes occur too frequently, organisms may have insufficient time to adapt to ecological space that becomes vacant. Sheldon (1992) suggested that stasis will therefore prevail under these conditions. If fluctuations persist, lineages may be favoured that can cope with a wide range of environmental conditions. These 'generalists' are more likely to persist than lineages of 'specialists' that are adapted to specific conditions but are more likely to go extinct when conditions change. In addition, fluctuating conditions may favour species that can track wide fluctuations by changing their distribution rather than evolving, because there is insufficient time for adaptation prior to the next environmental change.

We can relate this argument to the bet-hedging phenomenon discussed in the previous chapter, where it was shown that a generalist phenotype could be favoured over specialist phenotypes when conditions fluctuate. However, here we are considering environmental fluctuations that occur over a much longer time-scale of hundreds or thousands of years, and we are concerned with avoidance of extinction rather than fitness differences. As mentioned above, generalist lineages are more likely to survive mass extinction events. This suggests that the persistence of some organisms can be related to environmental changes occurring on a long time-scale.

Finally, we should mention the possible impact of stressful periods on morphological variability. As discussed in the context of extinctions, stressful conditions can trigger the production of evolutionary novelties by inducing morphological variability or by altering the fitness landscape. These might occasionally allow organisms to adapt to new conditions, leading to a major evolutionary change. Stressful periods can coincide with the appearance of new taxa. For instance, in 13 fossilized fresh water mollusc lineages in East Africa, Williamson (1981) found that a long period of stasis was interrupted by fossil beds in which relatively rapid changes in shell shape occurred. These changes coincided with times when there may have been environmental stresses affecting all lineages in parallel.

2. Explanations involving speciation

The second type of explanation for associations between environments and evolutionary patterns links environmental conditions with speciation. In the paleontological literature, speciation is often assumed to be associated with morphological changes, particularly when these changes follow a pattern of punctuated equilibria (see Eldredge & Gould, 1972; Stanley, 1979). It has been proposed that environmental changes leading to disturbed and stressful conditions may facilitate speciation events. To consider the validity of this proposal, we need to consider two questions. Firstly, are conditions where evolutionary novelties arise associated with an increased rate of speciation? Secondly, does speciation result in morphological changes, or are these just as likely to occur in the absence of speciation?

There is little doubt that speciation rates will vary in different environments. Although there are numerous models of speciation, the most widely accepted ones (known as 'allopatric' models) incorporate some component of geographic isolation among populations (Barton, 1989). Once populations become isolated, they will tend to evolve separately and diverge genetically. This divergence can occur for two reasons. First, isolated populations will be exposed to different selection pressures. As a result, different genes will be favoured in different populations, causing genetic divergence. Second, genetic divergence can occur because of genetic drift. The effects of genetic drift will increase when the population size is small.

Reproductive isolation is viewed as arising, at least initially, as an incidental effect of genetic divergence between geographically-separated populations. As genetic changes take place, some of them may involve loci contributing to isolation. Allele frequencies at these loci may be altered randomly because of drift, or because there are pleiotropic interactions between reproductive isolation and traits under natural selection. For instance, early reproduction might be favoured in one environment because food is abundant in spring, while late reproduction might be favoured in another environment because food is abundant in late summer. Isolated populations in these environments are expected to diverge in their timing of reproduction, and this may eventually be sufficient to produce reproductive isolation between them even when geographic barriers are no longer present.

It seems likely that the effects of both drift and selection will be enhanced in environments that are disturbed and that occasionally undergo major changes. When species encounter stressful conditions, populations will become reduced in size and some may become extinct. This can result in geographic barriers developing between relict populations separated by a habitat that was previously favourable. Reproductive isolation could develop more rapidly if relict populations are small, because this will enhance genetic divergence due to genetic drift. In addition, the availability of vacant ecological space could facilitate the development of isolation because populations moving into vacant space will be exposed to new selection pressures. Environmental changes resulting in the fragmentation of populations could therefore increase speciation rates by enhancing rates of genetic divergence. We should emphasize that this depends on populations remaining geographically isolated for many generations to allow sufficient time for divergence to occur.

The presence of vacant ecological space is thought to have contributed to the high rates of speciation on some oceanic islands. For example, in finches from the Galápagos Islands, speciation is thought to have been initiated when birds reached neighbouring islands with vacant space (Grant, 1986). As populations were established on new islands, they adapted to local conditions and remained isolated for a long time because migrants from neighbouring islands were rare. Some researchers have argued that the colonization of new islands by only a few individuals, followed by a rapid increase in population size, might further enhance the rate at which reproductive isolation in island populations develops (Carson & Templeton, 1984). However, this issue remains contentious because it is not clear how these colonizing or 'founder' events could cause genetic changes leading to isolation (Barton, 1989). Recently obtained experimental data do not generally support the notion that drastic changes in population size increase the rate at which reproductive isolation develops (eg., Moya, Galiana & Ayala, 1995).

While we can provide several arguments in support of high speciation rates when environmental changes lead to the fragmentation of habitats and populations, this does not mean that the speciation process itself results in morphological change. A major problem in assessing the association between speciation and morphological change is that there is often no independent criterion for identifying species. In the fossil record, species are invariably identified by morphological differences.

There is no doubt that speciation can occur without much morphological change. Species living today provide many examples of this. For instance, Larson (1989) examined species in a family of lungless salamanders. He identified 15 speciation events, but only three were associated with detectable morphological changes. This indicates that evolutionary stasis assessed on morphological criteria can occur in spite of speciation. Larson also found skeletal polymorphisms within populations, suggesting that evolutionary changes in the skeletons of salamanders are possible in the absence of speciation. Several other lines of evidence, summarized in Arthur (1984) and Levinton (1988), indicate that speciation events are not necessarily linked closely to morphological change.

Patterns of evolution with the features of stasis and punctuation can just as easily occur within evolutionary lineages in the absence of speciation. If a morphological trait is under strong stabilizing selection (ie., a particular optimum value of the trait is always favoured), the trait may not change much as long as the same selection pressures persist. In addition, a trait may show stasis when the selection pressures acting on it are continually changing in opposing directions, because morphological changes occurring over a short time would not be detected in the fossil record. Morphological changes that appear to be rapid in the fossil record could simply be the result of selection within a population in response to a changing environment. The examples in Chapter 3 indicate that evolutionary changes in a population can occur within a few thousand years or less, so punctuations in the fossil record do not necessarily require a speciation event.

Patterns of punctuated equilibria can even occur when traits are under continuous directional selection (ie. extreme trait values at one end of a distribution are always favoured). There may be periods of minor changes interspersed with periods of rapid changes as mutations or recombinants arise enabling favourable gene combinations to be formed (Parsons, 1983b; Johnson, Lenski & Hoppensteadt, 1995). For instance, Figure 6.10 shows changes in the number of bristles on the scutellum of *D. melanogaster* as a consequence of directional selection for increased bristle number. Periods of rapid change occur as major genes rapidly increase in frequency, and this is interspersed by periods of relatively little change. Although this example involves a short period of continuous selection in a laboratory environment, it is possible that these types of changes also occur in the fossil record over a long time span. There are several examples where discontinuities in the fossil record have been detected within the same species, at least when species are defined by morphological criteria. For instance in the Antarctic radiolarian, *Pseudocubus vema*, mean thoracic width increased rapidly in relatively brief phases within a time period of about 2.5 million years (Figure 6.11).

Despite these difficulties, there are still other ways in which speciation can enhance the rate of evolution. One possibility is Futuyma's (1989) suggestion that evolution may allow geographic differences that have developed between populations to be maintained. Although populations can develop genetic differences as they adapt to local conditions, these are likely to be transient for

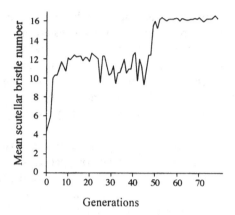

Figure 6.10. The response to directional selection for scutellar bristle number in *D. melanogaster*. The data are based upon females over 75 generations of selection. Note the two periods of rapid response lasting for very few generations in each case and the long stasis phase. (Modified in Parsons, 1983b from data of MacBean, McKenzie & Parsons).

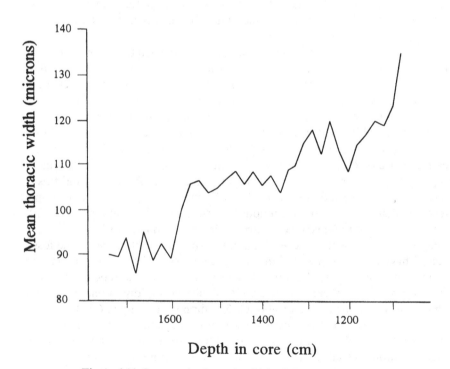

Figure 6.11. Increase in thoracic width of the Antarctic radiolarian, *Pseudocubus vema*, during an interval of about 2.5 million years. (Simplified from Kellogg, 1975).

two reasons. First, local environments will often change, so the same characteristics will not always be favoured in a population. Any genotypes that have evolved in response to one set of conditions may therefore be lost as populations adapt to new conditions. Second, a continual exchange of migrants between populations will result in the transfer of genes and reduce genetic differences between populations. However, gene flow between populations will be reduced once reproductive isolation develops, and any differences that may have developed between populations have an increased chance of persisting. Speciation may therefore enhance morphological change by decreasing the impact of gene flow. Speciation may not generate morphological variation, but may help to maintain it once variation develops between populations.

Another possible role for speciation is that it can result in evolutionary divergence between newly arisen species when they encounter each other and compete for similar resources. In this situation, members of a species that can avoid competition with the other species will have an advantage. The amount of overlap in the types of environments occupied by two species should therefore decrease over time. This process is thought to be important in morphological divergence among the Galápagos finches (Grant 1986). For instance, on one of the Galápagos islands (Pinta), the finch, *Geospiza fuliginosa*, feeds on seeds and nectar while another finch, *G. difficilis*, feeds largely on invertebrates in the leaf litter. *G. fuliginosa* is absent from another island (Genovesa), but here *G. difficilis* has a diet similar to *G. fuliginosa* even though the same food types are present as on Pinta (Figure 6.12). This suggests that interspecific competition influences the diet of *G. difficilis*. As a consequence, differences in bill form and body size have developed between populations of *G. difficilis* from Pinta and Genovesa. The morphological divergence between the two finches on Pinta would not have developed in the absence of reproductive isolation between the species.

In summary, morphological stasis may occur in stressed environments because of reduced biotic interactions or because harsh conditions make it unlikely that novel phenotypes will survive. Rapid evolutionary changes may take place because a sudden removal of stresses opens up new habitats. While this process may be more likely in disturbed environments, speciation does not appear necessary for large morphological changes, since rapid responses to environmental changes within lineages may also account for bursts of evolution. However, the development of reproductive barriers may help to maintain population differences and interspecific competition can further enhance morphological divergence once speciation has taken place.

We should point out that the above explanations are not necessarily mutually exclusive. Many instances of morphological evolution are likely to involve combinations of processes. For instance, novel phenotypes generated by stress exposure may only become established when vacant ecological niches are available. Once this leads to phenotypic differences between populations, the

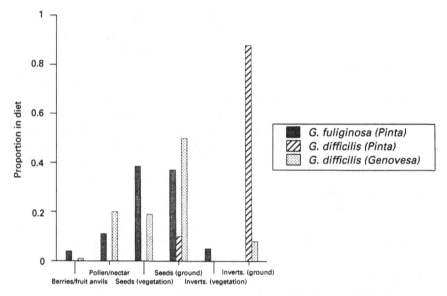

Figure 6.12. Diets of the finches, *Geospisa difficilis* and *G. fuliginosa* at lowland sites on two Galapagos islands (Pinta, Genovesa) during the dry season. *G. fulignosa* does not occur on Genovesa. (Simplified from Grant, 1986).

persistence of these differences may depend on the development of reproductive barriers.

Genetic changes underlying microevolution and macroevolution

Closely tied to issues raised in the previous section is the question of whether or not the types of evolutionary changes in the fossil record (macroevolutionary changes) can be related to evolutionary processes observed in extant populations (microevolutionary changes). Do the same types of genetic changes underlie these two processes, or are they in some way different?

This question is related to the way researchers envisage evolution occurring. Darwin (1859) often argued that evolution by natural selection proceeded via the accumulation of successive favourable variations that had small effects on a trait, producing no great or sudden phenotypic shift, but giving a gradualistic pattern of change. This pattern was envisaged as arising from microevolutionary processes, such as those discussed in Chapter 3. It is based upon the tenet held by many evolutionary biologists that adaptations normally involve the substitution of many genes each having small effects. However, as we have already mentioned, many evolutionary changes often do not fit this pattern, but occur in a discontinuous fashion. It has been argued by Eldredge & Gould (1972) and others that major evolutionary changes, such as those that occur during punctuations, may involve different genes than those that normally contribute to genetic variation within and between populations. These sudden changes are thought to be associated with genes having very large phenotypic effects.

Because the fossil record usually monitors changes in morphology, we need to consider genes underlying morphological traits in examining this question. There are two issues that we need to address, namely the number of genes involved in major morphological changes and the nature of these genes.

Genetic data on the number of genes controlling traits has been sought by undertaking crosses between strains of the same species and crosses between related species. Many crosses have been carried out between strains of animals and plants that differ for morphological traits after they have been artificially selected. Crosses have also been carried out between different varieties of plants or breeds of animals that differ in morphology. These crosses tend to indicate that morphological differences between strains or varieties involve several genes, although some genes often have quite large effects on traits, accounting for more than 10–20% of the differences between selected lines or varieties. Thus genetic changes in morphological traits generated by artificial selection are likely to involve several genes rather than just one or two, although there is not much evidence that numerous ($\gg$10) genes with small effects are involved (Mackay, 1995).

Unfortunately, these types of comparisons may not be relevant to natural populations, where population sizes are larger and where selection may be of a lower intensity, and certainly less directed, than imposed by artificial selection. We have previously discussed the potential effects that selection intensity and population size can have on estimates of the number of genes influencing a trait (Chapter 3).

There are a few cases where morphological shifts in natural populations have been analyzed genetically, and these are reviewed in Orr & Coyne (1992). In some cases, such as in the evolution of mimicry in butterflies, morphological shifts appear to involve major genes. When mimicry develops, genetic changes in the major genes of a palatable species allow it to resemble a distasteful species. In other cases, several genes may be responsible. For instance, crosses between related *Drosophila* species with a different head width indicate that several genes are responsible (Templeton, 1977). However, it is not clear if 'several' genes means numerous genes with small effects or a more restricted number of genes. Unfortunately, most of the genetic studies on morphological differences between populations and species are flawed in some way (Orr & Coyne, 1992). We are therefore left with the impression that genes having large effects can be important in morphological divergence between species and within species, but their relative importance compared to minor genes is unknown. Even if we had a good estimate of the number of genes involved, we would still not know if all of these genes were actually involved in responses to environmental changes in geological time. As discussed in Chapter 5, genetic differences in a trait between populations or related species can continue to develop even when the trait is no longer under direct selection.

While research on the number of genes may provide some insight into differences between microevolution and macroevolution, we need to identify specific genes to directly compare the types of genetic changes that are important at these levels. Unfortunately, genes underlying morphological variation within species and populations have generally not been associated with those accounting for species differences.

It is possible that genes associated with major morphological changes are in some way different to those segregating within populations (Arthur, 1984; McDonald, 1990). Many genes contributing to variation within populations regulate the expression of enzymes, and most of these may have minor effects on morphological traits. However, other genes with major effects on morphology can affect the development of organisms.

Major advances have recently been made in understanding genes controlling development in *Drosophila*. The process is initiated by a series of genes that are encoded in the mother's genotype. Products of these maternal genes determine the way cells are arranged inside the oocyte. Genes in the zygote are then activated, and these further partition the embryo and determine a segmentation pattern in the embryo. This segmentation pattern persists in the larval and adult stages. Segmentation is followed by the activation of another set of genes, the homoeotic genes or *Hox* genes first mentioned in Chapter 2, that determine the organization of each segment. A further set of genes controls differentiation in areas within segments. The genetic control of development in *Drosophila* therefore involves a cascade of genetic actions.

Mutations in genes that act early in development tend to be lethal and are therefore unlikely to be important in evolution. However, genes acting later in development could mediate major morphological changes. In particular, homoeotic mutations can cause one part of the body to be transformed into another. Such mutations can mimic many major morphological differences among groups of insects (Carroll, 1995). For instance, the dramatic differences in the hindwings of flies, beetles and butterflies appear to involve *Hox* genes. The existence of developmental genes with large effects on morphology is also apparent from the major morphological changes that can be selected when *Drosophila* and other organisms are exposed to stresses at specific developmental stages (Chapter 2).

Hox genes occur in other animal groups although their mode of action may differ from that in insects (Carroll, 1995; Williams & Nagy, 1995). For instance, in vertebrates these genes control the development of vertebrae, limbs and the central nervous system. A duplication of the cluster of *Hox* genes distinguishes primitive chordates from vertebrates, and this duplication may underlie the evolution of many of the complex features of the vertebrate body plan.

Arthur (1984) and others have argued that major evolutionary shifts are unlikely to occur without developmental genes such as the homoeotics being involved. The number of genes having these major developmental effects is likely to be limited. It therefore seems unlikely that developmental differences between species could be accounted for solely by changes in minor genes. Nevertheless it is difficult to test this idea directly by crosses between species. It is difficult enough crossing closely-related species with the same developmental pattern, let alone species differing markedly in their development.

Recent advances in molecular genetics may help to elucidate the nature of genes involved in macroevolutionary changes. McDonald (1990) suggested that developmental mutations resulting in morphological change may be associated with transposable elements. These elements are common in the genome, and their induction can cause mutations in genes regulating enzymes and controlling development. Moreover, transposable elements can be induced to insert into genes and cause mutations by environmental stress (Chapter 2). As we have already noted, periods of intermittent stress tend to be associated with major morphological changes. McDonald (1990) suggested that this is more likely to occur when populations are small. Normally, populations contain alleles that supress transposition. However, these alleles may be lost by chance in small populations because of genetic drift. Thus rapid environmental changes could indirectly induce transposition events by reducing population size, as well as more directly via the effects of stress. The involvement of transposons in regulatory evolution should become clearer as the molecular basis of genetic variation in developmental genes is examined.

Responses to climatic change: evolution versus distribution changes

The above findings indicate that morphological changes in the fossil record can often be associated with periods of environmental change, but they tell us little about the potential of species to adapt to changes once they arise. Obviously, organisms have been unable to adapt to the major global changes that underlie mass extinctions, but what about other environmental changes? Do species usually go extinct locally, or are they able to adapt partially or completely to counter such changes?

The fact that many species have become extinct outside times of mass extinction indicates that they do not always successfully adapt to environmental changes. A lack of adaptation is also evident from documented shifts in species distributions as conditions change. The fossil record is replete with species that are now extinct, but have undergone major distributional shifts in the past in response to climatic change. There are also many surviving species that have undergone documented geographic shifts, particularly during the climatic changes of the last 100000 years. These changes will often reflect a tendency for species to track favourable ecological conditions without adaptation. For instance, the distribution of animals adapted to warm environments normally shrinks during periods of cooling, and expands during periods of warming.

Distribution shifts can depend on factors not directly related to the climatic conditions themselves. The expansion of species into favourable areas will depend on their ability to locate these areas. For instance, populations of trees often expand at a much slower rate than predicted for climatic changes, indicating that dispersal is inadequate to keep up with environmental changes. Distribution changes can also be influenced by biotic interactions among species; the expansion of one species can be limited by the presence of another species that is less well-adapted to the environmental conditions, but is nevertheless able to exclude it. Finally, distribution changes in parasites and predators will depend on their prey or hosts, rather than directly on climate.

Because migration and biotic interactions will influence distribution shifts as well as adaptation, changes in the distribution of a particular species cannot be used to evaluate the evolutionary potential of this species. However, we can infer the absence of evolutionary change by looking at a group of coexisting organisms. Species within a group may undergo similar changes in their distributions, or else their distributions may change individually and largely independently of each other. If the former occurs, adaptation has probably not occurred because most species in a group are unlikely to have the same evolutionary potential (Chapter 4). However, if the distributions of related species change individually, there is the possibility that they have adapted to different extents.

Relevant data are provided by Coope's (1979) study of Coleoptera (beetle) fossils from the Quaternary in Europe. Only a few species of beetles showed morphological changes in this period, suggesting little evolution in morphological traits. In contrast, there were marked changes in the distribution of beetle species. While physiological evolution cannot be excluded by morphological stasis, Coope suggested that inferences about physiological changes can be made by investigating assemblages of Coleopteran species. Importantly, beetle species found together in environments today are generally the same as those that occurred together as fossils. There were some exceptions where species were found in different associations, but these represented less than 5% of the total number of beetle species. This suggests distribution changes without large scale physiological evolution, because it seems unlikely that all

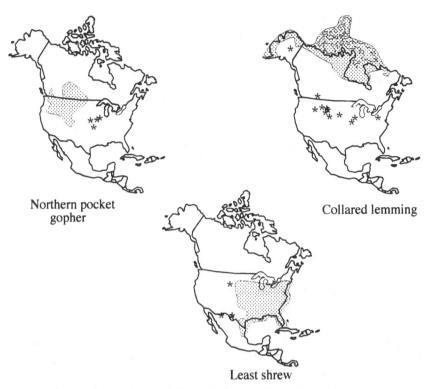

Northern pocket
gopher

Collared lemming

Least shrew

Figure 6.13. Modern distribution of three small mammals in North
America, along with fossil localities of the same species in the late
Wisconian (indicated by asterisks). (Fossil localities mark the eastern,
western and southern extension of the gopher, shrew and lemming
respectively. (After Graham, 1992).

beetles in an assemblage show the same ability to adapt to environmental
changes.

   While beetles seem to have maintained the same species associations as their
distributions changed, this is not generally true for other groups of organisms.
For instance, Graham (1992) described the case of three small mammals
which had overlapping ranges in North America, but which experienced
rather different changes in distribution during the late Pleistocene. As evident
from Figure 6.13, one of the species contracted westward, the other moved
eastward, and a third species became restricted to the northern part of North
America. Individual shifts in distribution have been described in many other
groups of organisms. In particular, trees which grow together today have
usually been associated with other species of trees in the Quaternary. Even
many insects appear to have responded to environmental changes on an individual basis rather than as an assemblage (e.g. Downes & Kavanaugh, 1988).

This common tendency for species to change their distribution on an individual basis does not prove that physiological adaptation is occurring to a variable extent in different species. It is always possible that related species differ in their dispersal ability or that they have been influenced by different biotic interactions. Nevertheless, the fact that individual responses by species seem to be the rule for many groups of organisms suggests that physiological adaptation could be common. When organisms encounter novel conditions, physiological changes may initially be selected, prior to morphological changes. For instance, Fisher (1977) showed that clones of diatoms from fluctuating estuarine environments are much less sensitive to novel chemical stresses than clones from more stable ocean environments that are morphologically identical. Novel stresses would select for estuarine clones and cause a physiological change without morphological evolution.

Summary

Environmental changes have occurred throughout geological time. Periods of extremely stressful conditions on a global scale are associated with mass extinction events detectable in the fossil record. Many taxa disappear during such events, and they are followed by periods of rapid diversification. In this sense, some have argued that mass extinctions can indirectly trigger evolutionary change.

Widespread taxa in the fossil record and small organisms appear to be relatively less susceptible than restricted and larger organisms to environmental changes resulting in extinction. Some organisms may have a reduced susceptibility to extinction because of stress evasion mechanisms including stress-resistant life-cycle stages with low metabolic rates. Insects and especially plants may also be less extinction prone because of resistant life-cycle stages.

Evolutionary novelties appear to be mainly a feature of disturbed habitats with high levels of primary productivity. When productivity is severely restricted, little evolutionary change occurs at the morphological level, as in the case of 'living fossils' in fluctuating environments. Evolutionary patterns involving stasis interspersed with large, rapid morphological changes may be characteristic of environments that fluctuate over thousands of years and where major environmental shifts take place occasionally. Gradual changes in the fossil record are likely to be a feature of moderately stressed environments that fluctuate less over a long time-scale. When fluctuations are minor, relict species occur.

There are a number of explanations for associations between evolutionary rates and environmental conditions. A burst of evolution may be triggered by vacant space generated following extreme environmental changes that reduce biotic interactions and allow the survival of novel phenotypes. Extreme conditions may also induce evolution by generating variability and altering fitness landscapes. Morphological evolution is not necessarily associated with speciation events, although speciation may have a secondary role in morphological change.

Rapid morphological changes may involve a few major genes and could be associated with transposable elements acting on regulatory genes. Species have often shifted their distributions in response to environmental changes. Species change distributions individually rather than as assemblages, which may reflect differences in migration potential, biological interactions, or physiological adaptation.

CHAPTER 7

Conservation and future environmental change

In the above chapters, we have investigated how organisms have adapted to environmental changes in the past. We have also considered some of the factors that can influence their ability to adapt. Here we consider these findings in the context of the potential for adaptation to future environmental changes. Can we predict the ability of organisms to adapt? What strategies can be implemented to maximise this potential?

We start by briefly discussing the types of environmental stresses organisms may encounter in the future, and look at the biological changes already being detected in vulnerable habitats. We then examine the ways in which genes increasing adaptive responses to these changes might persist in populations, particularly in the face of increasing fragmentation of the habitat of organisms. The possibility of incorporating some of these considerations into management programmes for endangered species will be assessed.

Future environments: how will organisms need to adapt?

While it is difficult to predict future climatic trends, most models of global climate envisage increases in world temperature for the next century. Compared with prehistoric changes of a similar magnitude, the predicted increase is likely to be exceedingly rapid. Estimates vary, but are typically around 1°C above the present temperature by 2025AD and 3°C before the end of the next century, or 2–4°C above that of the preindustrial period (Schneider, 1993). The impact of the increase in global temperature is likely to vary considerably between regions. For instance, the central areas of continents are expected to change more rapidly than coastal regions, and temperature changes in the tropics are expected to be less than at high latitudes.

Changes in precipitation patterns are expected to occur along with those involving global temperature. Average precipitation and evaporation should increase, although this may vary between regions. At high latitudes, increases in precipitation are predicted throughout the year, but at lower latitudes changes are more unpredictable. Some climate models even predict decreases in precipitation on some continents.

The effects of global changes will be exacerbated by more localized human activities, particularly those involving the destruction of natural habitats. This is evident from the effects that human activities are having on local climate. In the Amazonian tropics, rapid deforestation of rainforests is leading to significant increases in surface temperature, a decrease in evapotranspiration and precipitation, and an extended dry season. A model on the effects of replacing forests by pasture (Shukla, Noble & Sellers, 1990) indicates that deforestation may eventually have a substantial impact (Figure 7.1). For instance, surface and soil temperatures could increase by 1–3°C in the deforested regions, a shift matching the predicted global climate changes in the next century.

Because of these global and local effects, many organisms are likely to come under stress related to temperature extremes. It is widely accepted that an increase of a few degrees can have a major impact on the distribution and abundance of species. Potential effects can be readily predicted for ectotherms whose activity is restricted to a narrow range of temperatures. For example, in a population of the lizard, *Sceloporus merriami*, Dunham (1993) estimated that a 2°C rise in average air temperature would reduce the time above 32.2°C when lizards become active from 2.5 hrs to 1.92 hrs per day, a fall of 23%; for a 5°C rise in air temperature, activity time would be reduced to 0.58 hrs, a fall of 76.8%. Reduced activity will in turn diminish the available time for foraging and for social interactions (courtship, territorial defence etc.). In addition, lizards with higher body temperatures will have higher maintenance costs since metabolic rates increase with temperature. It is therefore unlikely that this species will be able to withstand a change of a few degrees in the absence of adaptation to warmer temperatures. Many other ectotherms are also unlikely to survive changes of a few degrees if they do not adapt to the new conditions.

The immediate effects of temperature changes will be less evident in endotherms because they are able to compensate at least partially for altered external conditions. However, the survival of endotherms can also be directly related to environmental temperatures. In fasting rats, heat production is at a minimum between 28–29°C, and increases at lower temperatures and to a lesser extent at higher temperatures (Kleiber, 1961). As the temperature is reduced or increased, the survival time of starving rats becomes progressively shorter. Similar effects are found in other small mammals such as mice and guinea pigs (Blaxter, 1989). Relatively minor changes in environmental temperature can therefore increase mortality in small endotherms.

Temperature changes can also affect organisms indirectly by influencing the timing of life cycles, particularly in the tropics where dormancy can be determined by very small changes. For instance in a ground beetle, *Africobatus harpaloides*, temperature variation controls dormancy despite the fact that the mean maximum temperature throughout the year in its Central African forest

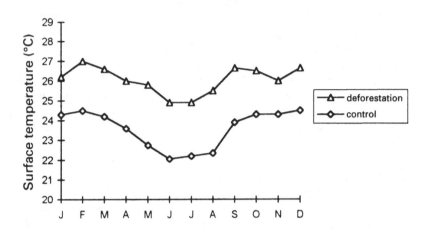

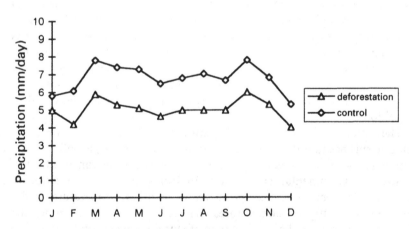

Figure 7.1. Predicted effects of the conversion of Amazon rainforests to pasture on monthly surface temperatures and precipitation in this region. (Simplified from Shukla *et al.*, 1990).

habitat varies by only 0.9°C (Greenwood, 1987). Slight increases in temperature could eliminate dormancy in such species and lead to extinction.

Many organisms rely on seasonal temperature changes to time their reproduction at periods of abundant food. This process can be upset by climatic changes, leading to reproductive failure. An example of this possible scenario is provided by Myers & Lester (1992) who considered the effects of changing climatic conditions on migration patterns of shorebirds. These birds are obligate migrants because of their need to track food resources. Many shorebirds travel several thousand kilometres annually as they move between breeding grounds at high latitudes to overwintering grounds in the tropics or subtropics. During this migration, shorebirds rely on the availability of transient resources. For instance, in Delaware Bay in North America, shorebirds rapidly replenish energy and protein reserves by feeding on the eggs of horseshoe crabs that are deposited in a brief interval, before moving to the arctic and feeding on a short bloom of insects during the breeding season. Both the availability of insects and crab eggs are determined by climatic factors. If birds fail to encounter these resource flushes, they will be unable to complete their migration and breed in the arctic. Myers & Lester (1992) suggested that global warming will have a particularly serious effect on these animals because of the likely uneven changes in temperature. Global temperature changes are predicted to be greater at high latitudes than at lower ones. Hence blooms of resources such as crab eggs along migration routes will no longer be separated by the same time interval from blooms of arctic insects. The latter may bloom several weeks earlier, whereas the peak availability of crab eggs may only change by a few days, resulting in starvation and breeding failure for the shorebirds.

The above effects are all related to changes in the mean temperatures of environments. We also need to consider the effects of changes in extreme conditions. Many organisms are particularly susceptible to short stressful conditions at one or more of their life cycle stages. In animals, the young tend to be more susceptible to temperature stress than adults. For instance, adult birds can tolerate 43°C for long periods of time, but these conditions tend to be lethal for the early developmental stages of avian embryos (Dawson, 1992). In our own species, the newborn have a metabolic rate far higher than adults and they are highly sensitive to a range of stresses including starvation, dehydration, disease and climatic extremes. In plants, some developmental stages can be particularly sensitive to stressful periods. For instance, extremes of heat, cold and drought can have large effects at critical times on the yield of agricultural crops (Katz & Brown, 1992). In corn, runs of five consecutive days of extreme heat during the temperature-sensitive silking stage can result in complete crop failure (Mearns, Katz & Schneider, 1984). When predicting effects of climatic changes, the impact of extremes as well as average conditions needs to be considered.

Any biological effects of temperature changes on individual species will be compounded by inputs from other stresses. When organisms under one type of stress are exposed to a second stress, they can suffer additional deleterious effects. For instance, Feder & Shrier (1990) considered the effects of ozone and ultraviolet radiation (UV-B) on two plants, *Nicotiana tabacum* and *Petunia hybrida*. Ozone is a common pollutant produced by emissions from vehicles and power plants. Stress effects were measured by examining the growth of pollen tubes, because this trait is essential for fertilization and because it is influenced by UV-B and ozone. The data (Figure 7.2) indicate that tube lengths are reduced to a greater extent when there are two stresses present than when there is only a single stress. Another example involves amphibians exposed to UV-B and pH. In the frog, *Rana pipiens*, Long *et al.* (1995) found that survival of embryos was decreased only when low pH was combined with UV-B radiation and not when either stress was present on its own.

The importance of interactions between stresses has also been emphasized in explanations for the decline of forests in Europe and North America (Hinrichsen, 1986). Air pollution has resulted in the deposition of toxins and substances that can influence the growth of trees. These are often considered as the primary causes of forest decline, resulting in a decrease in the rate of photosynthesis and the accumulation of toxins. Because the energy balance of trees is affected, trees have fewer resources to counter the effects of other stresses. They therefore become more susceptible to climatic extremes such as drought and frost, and are more prone to diseases. Forest decline may also be linked to soil acidification resulting from pollution. Acidification can cause the loss of nutrients and the release of toxic aluminium in soil. The roots of trees become weakened and may then become susceptible to other stresses such as drought. Any climatic changes arising from global warming are therefore likely to accentuate the effects of atmospheric pollutants on forest decline.

While we have so far focussed on the effects of environmental changes on individual species, many additional effects are likely to result from interactions among species. In Chapter 1, we discussed ways in which changes in environmental conditions can have numerous secondary effects, particularly by influencing biotic interactions among species. Global warming is likely to have these effects as well, by altering competitive interactions, predation patterns and the incidence of parasites and pathogens. Much of the concern about global warming has focussed on likely effects of pests and pathogens increasing in abundance and expanding their geographic ranges. Since cold winters exert a restraining influence on many insect pests, milder winters from greenhouse warming may lead to explosions of pest numbers. Many parasites and diseases restricted to tropical and subtropical environments are expected to become widely distributed as global warming takes place. For example a 1°C rise in average temperature could mean a substantial range expansion of the tropical tse tse fly which carries the parasite causing sleeping sickness in man.

In summary, increased global temperatures and changing rainfall patterns may exert a direct effect on organisms by causing high temperature stress and restricting the ability of organisms to reproduce. Temperature changes may also have more subtle effects by altering the timing of different

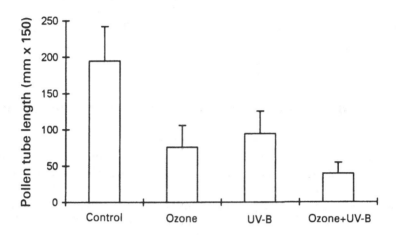

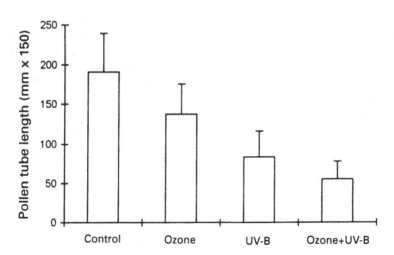

Figure 7.2. Effects of ozone, ultraviolet radiation and a combination of these stresses on the growth of pollen tubes in *Nicotiana tabacum* and *Petunia hybrida*. (Plotted from data in Feder & Shrier, 1990).

developmental stages, and by influencing the availability of resources for reproduction. Interactions are likely between temperature, drought, pollutants and other stresses. These will complicate the types of environmental changes to which organisms have to adapt. Direct effects on organisms will be coupled to a cascade of indirect effects since the distribution and abundance of food sources, predators, parasites and pollinators may be altered.

Biological monitoring: evolutionary considerations

The above section illustrates the types of environmental changes organisms are likely to encounter in the future. How will the effects of such changes be detected, particularly when species can counter them by moving away or by evolving resistance? We considered ways of detecting stressful conditions in Chapter 1. These included monitoring changes in biochemical parameters such as AECs, as well as monitoring traits related to fitness. These measures may be useful in detecting when environmental changes are likely to have an impact on natural communities.

There are two ways organisms can be used to monitor changes within an area. The first considers organisms that already exist in an area. Changes are monitored by following the abundance and distribution of individual species, although variation within a species can also be monitored. The second way is to focus on individuals known to be susceptible to particular environmental stresses. This is a common approach for examining the effects of chemical pollutants. Susceptible animals are released or held in cages within an area, while susceptible plants are established in test plots. Alternatively, the effects of known levels of stresses in an area are tested by exposing susceptible organisms under laboratory conditions.

A number of stress-sensitive organisms may be useful for the first approach, including lichens, forest fungi, amphibians, fish, small invertebrates and many plants. In Chapter 6, we discussed how lineages with characteristics such as widespread distributions and stress-resistant stages in their life cycles were relatively less prone to extinction during mass extinction events. These species are also likely to be the least affected by global climatic changes. Conversely, some species will be particularly susceptible to the effects of climatic changes, and these may serve as biological monitors of stressful conditions. Such species will be particularly useful if they are unable to rapidly evolve mechanisms for resisting or evading changing conditions.

Ectomycorrhizal fungi are useful for monitoring changes in forests (Fellner, & Peskova, 1995). In Europe, there has been a recent decline in the abundance and diversity of these organisms whose presence is detected by above ground fruiting bodies. This decline may reflect changes in levels of nitrogen, sulphur and ozone in the air. Because the threatened fungi mainly live in close symbiotic associations with trees, providing trees with water and minerals in exchange for carbohydrates, a mass extinction of fungi can lead to a decline in the health of trees, which will be accentuated if trees are also exposed to other stresses.

Another group of organisms that is useful in this context is the amphibia. These possess several characteristics that make them susceptible to pollutants and other stresses. Following Dunson, Wyman & Corbett (1992), amphibians have features likely to make them sensitive when exposed to a range of pollutants, such as: (1) complex life cycles incorporating both aquatic and terrestrial stages, resulting in exposure to pollutants in both habitats; (2) permeable eggs, gills and skin enabling rapid absorption of toxic substances; (3) poikilothermy making them vulnerable to changes in environmental temperature; and (4) exposure to toxic substances especially during aestivation and hibernation.

Amphibia appear to be declining on a global scale. For instance, populations of several species of frogs that inhabit relatively pristine areas of western North America have declined drastically (Blaustein & Wake, 1990). However, this is not a feature of all amphibian groups or all localities, and it can be difficult to separate natural changes in population sizes from human activities. Extinctions of frog populations are occurring in regions that appear protected from human influences, suggesting that direct effects of local human activities are not always involved. One possible explanation is that declines are a consequence of UV-B radiation which kills the eggs of some amphibians but not others (Blaustein *et al.*, 1994) and may interact with low pH stress as mentioned earlier. If there are major world-wide declines of amphibian populations, there will be an impact on other organisms. This is because amphibians are an important component of many ecosystems, and often represent the largest vertebrate group in terms of biomass.

For any group of organisms to be useful as monitors of environmental deterioration, natural fluctuations in populations need to be distinguished from those induced by environmental changes. Natural fluctuations will occur if there are changes in factors such as food resources, parasitism or predation rates. One way of controlling for these types of factors is to consider a closely-related species that has evolved sufficiently to differ in its susceptibility to environmental changes. Closely-related species are likely to use similar food resources and have similar predators. If one species has a widespread distribution and a greater resistance to environmental stresses than a related species, changes in the relative abundance of both species can be used in monitoring stressful conditions. For instance, members of the genus *Drosophila* comprise species that are sensitive to temperature extremes and low humidity, as well as species that are resistant to these stresses (Chapter 5). In Queensland, Australia, many rare species are restricted to threatened pockets of rainforest that are moist and have a relatively small temperature range of around

15–22°C. As the temperature increases towards 25°C at the edge of rainforest pockets, there is a tendency to find only related species that are more wide-spread. Rare species can therefore be sensitive to climatic shifts comprising a temperature increase as small as 1–2°C, especially if there is an associated fall in humidity. Laboratory tests on some of the species confirm that rare species are sensitive to high temperature and desiccation compared with their common relatives (Parsons, 1992a). We could therefore expect rare species to become threatened as stress levels increase, leading to a gradual reduction in speci᷈s diversity.

Or᷈anisms such as amphibia and fungi are useful monitors of large scale environmental changes because their declines seem to occur over a wide area. The fact that abundance is declining suggests that these organisms have a limited ability to evolve in response to environmental changes. Presumably, the types of selection limits discussed in Chapter 4 prevent adaptation. This makes such organisms useful monitors, although it is always possible that the recovery of a population may reflect adaptation rather than the removal of a stress. The susceptibility of a population to a particular stress therefore should really be established before it can be used as a monitor for that stress.

In addition to monitoring environmental changes on the basis of species composition, we can use genetic variation within species as an indicator of stress. As changes occur, genotypes that are relatively more successful at resisting or evading stresses should increase in frequency in populations. Unfortunately, it is difficult to follow genetic changes in natural populations of organisms, particularly for loci likely to be involved in adaptive responses to environmental changes. A genetic approach becomes more feasible if loci under selection influence morphological traits. For instance, variation in wing colouration of butterflies and banding patterns of snails are often related to temperature selection (Chapter 3), and changes in the frequency of different morphs could indicate environmental changes. However, these types of poly-morphisms have only been described for a few organisms. The frequencies of morphs may also be influenced by selection pressures other than those directly associated with environmental changes. For instance, the banding morphs of snails and wing patterns of butterflies can both be influenced by predation as well as selection due to temperature extremes.

Phenotypic variability within species can also be monitored. We discussed several phenotypic changes associated with unfavourable conditions in Chapter 2. Among these, fluctuating asymmetry (FA) has been used exten-sively as a morphological monitor of stressful conditions and it has been sug-gested that FA provides a useful early warning system for identifying populations threatened by environmental stresses (Clarke, 1995). However, major effects on FA tend to become apparent under stresses close to the limit of survival of organisms. For this reason, FA is likely to be a useful monitor only in marginal habitats where stresses are severe.

We now turn to the approach of monitoring environmental changes by using specific organisms. These can either be introduced into an area, or tested in the laboratory using levels of pollutants known to exist in an area. Useful organisms need to be easy to culture and handle, as well as being susceptible to environmental stresses. For instance, the small crustacean Daphnia is often used to test for the effects of pollutants in freshwater, because Daphnia is susceptible to a range of chemicals and can be readily cultured in the laboratory. A number of cultivated plants such as tobacco and beans can measure effects of air pollutants, because they are susceptible to pollutants and can be grown easily and rapidly.

When devising ways of monitoring environmental changes, there is often a conflict between repeatability and ecological relevance (Forbes & Depledge, 1992; Calow, 1992). This conflict arises for two reasons. First, tests are likely to be highly repeatable if they are carried out under controlled conditions, particularly in the laboratory. However, these conditions may be far removed from nature where environments are complex and tend to fluctuate. Second, tests will be most repeatable if they are carried out on specific strains of organisms. For instance, inbred strains of rats or clones of Daphnia are typically used, and tests are often carried out with just a single strain or clone to ensure repeatability between different laboratories. However, the response of one strain or clone will often differ considerably from that of other conspecific individuals because of the ubiquity of genetic variation for stress responses (Chapters 2–4). Results obtained with one strain may therefore be difficult to extrapolate to an entire species.

Because of these complexities, recent work has emphasized the usefulness of carrying out tests with a range of strains or clones under natural conditions. For instance, Lovett Doust, Lovett Doust & Schmidt (1993) advocated using a range of genotypes from aquatic plants in testing toxins in water and sediments. Plants may be particularly effective because they can be grown at sites where environmental changes are suspected. Plants can also detect the cumulative effects of pollutants because they can be left in a substrate for a long time. Because plants can be propagated vegetatively, it is possible to carry out experiments at different sites with the same range of genotypes, thereby enabling assessment of genetic variation in responses to stresses.

Mutants and varieties that are particularly susceptible to stresses may also be useful as indicators of stressful situations. In *Drosophila melanogaster*, mutants such as 'shaker' that have high metabolic rates and are behaviourally active are highly sensitive to stresses such as elevated temperature, desiccation and an unsaturated aldehyde, acrolein. Such mutants can monitor air pollution levels (Parsons, 1992b). In plants, particular cultivars may be more sensitive to pollutants than others. For instance, Manning & Feder (1980) described the susceptibility of a particular cultivar of tobacco (Bel-W3) to ozone. This pollutant causes flecking on leaves, and the degree of flecking when Bel-W3 plants are placed in an area can provide a measure of ozone concentrations. Other varieties of tobacco that are resistant to ozone can act as controls. These procedures are quite sensitive. However, they can only be of use as early-warning indicators, because susceptible strains are unlikely to reflect responses in natural populations of the same species.

In summary, our growing concern with stress in the environment has highlighted the need to develop efficient and sensitive monitors of stress. Changes in the abundance and distribution of stress-sensitive species provide one indicator, particularly when related species that are more resistant to environmental changes are simultaneously monitored. Variation within a population can be used to follow effects of environmental changes. Genetic variation needs to be considered when developing strains for environmental monitoring. However, a conflict often occurs between using a range of strains to achieve relevance to natural populations, and the need for a single clone or strain to achieve repeatability in testing for stress effects. This implies a need for large and well-designed experiments to monitor stress effects.

The effects of future environmental changes: extinction or adaptation?

The extent to which populations can adapt to changing environmental conditions is ultimately limited. The fossil evidence paints a bleak picture of the ability of species to adapt to environmental changes because most species are extinct. The existence of stable species borders (Chapter 4) indicates that most species cannot mobilize underlying genetic variation to adapt to environmental conditions immediately beyond their borders, even though there are many examples of rapid evolutionary change within species (Chapter 3). This indicates a low probability of adaptation to rapid environmental changes arising from global warming and makes it likely that the earth today is in an escalating mass extinction event. However, as we noted in Chapter 6, it is possible that many species have undergone partial adaptation over geological time, as suggested by the fact that similar species generally do not show the same shifts in distribution following a climatic change.

To what extent can environmental changes be countered by adaptation? We have seen in Chapter 3 that species have the potential for evolutionary change, even though this potential is restricted because of tradeoffs and other factors. Can we predict the likelihood of adaptation despite these complications?

Interactions between factors likely to influence the potential for adaptation are summarized in Figure 7.3. The ability of species to adapt will depend on genetic variation, which, in turn, depends on migration causing an influx of new genes, and on the size of the population. Environmental changes can influence the presence of other species, and these can affect population size.

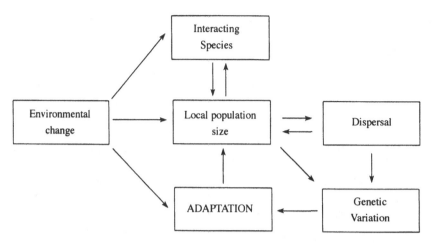

Figure 7.3. Interactions between factors likely to influence the response of a population to an environmental change. (Modified from Lynch & Lande, 1993).

For instance, predators, parasites and competitors can all decrease the size of populations.

The simplest way of predicting the potential for adaptation is to ignore effects of population size and biotic interactions in Figure 7.3, and consider a large population over a short time interval. If this population is under directional selection, its response to selection in the short term (R) is given by

$$R = h_n^2 S,$$

where h_n^2 is the narrow-sense heritability of a trait (see Chapter 2) and S is the 'selection differential'. S measures the extent to which the mean value of individuals that generate the next generation differs from the mean value of the overall population. The equation can be used to predict if a population will respond successfully to a changing environment. For instance, Billington & Pelham (1991) considered the time when birch trees first burst their buds in populations from Scotland. They calculated the heritability of this trait and then asked if the heritable variation was sufficient to allow birch populations to counter the effects of global warming. Billington & Pelham assumed a 2°C rise in temperature over 60 years, allowing for 12 generations of selection on birch. The 2°C rise is expected to advance budburst by 40 days, requiring an increase in budburst of 3.3 days per generation. The selection differential needed to produce this response can then be calculated. For instance, if h_n^2 is 0.59 and an increase of 3.3 days is needed per generation, then parents will need to have a budburst 0.33/0.59 or 4.6 days later than the parental mean. To see if this is possible, the phenotypes of individuals in the population were examined. Individuals that exceeded the value of 4.6 days existed at an appreciable frequency in only two of the ten populations that were examined. This suggests that most birch populations are unlikely to adapt to changing environmental conditions in the short-term.

Lynch & Lande (1993) viewed the problem in a different way. They considered a model where a trait is under stabilizing selection so that an optimum value is favoured. They examined the situation where the optimum shifts as the environment changes. As this happens, the mean of a population will lag behind the optimum until adaptation is complete. If the lag is large enough, the fitness of the population will decline, along with its size. This may lead to extinction of the population if its growth cannot be maintained (ie., the fitness of the population becomes negative). Lynch & Lande (1993) considered the rate of environmental change to which a population can adapt under this model; i.e., the rate of change that can be tolerated without population growth becoming negative.

In this case, the rate of population growth is defined as

$$r = r_m - \frac{(\bar{g} - \theta)^2 + V_P}{2\sigma_w^2}$$

where r is the rate, r_m is the rate of increase for the optimum phenotype, V_P is the phenotypic variance, θ is the optimum phenotype and σ_w is the width of the fitness function. We met the V_P term earlier in Chapter 2 and this variance is composed of the genotypic (V_G) and environmental (V_E) variances. The fitness function gives the fitness of an individual with a particular phenotype. As it increases, the difference in fitness between individuals with phenotypes that are a given number of units apart declines. The selection intensity therefore decreases as σ_w^2 increases. The population is doomed if r is less than zero for a long time.

In the equation, r will be less than r_m because a variety of genotypes are present and not all genotypes will have the optimal value θ. The presence of non-optimal genotypes imposes what is known as a 'genetic load' on the population. If the load becomes too great and genotypes differ too far from the optimal value, the population growth rate will fall below zero and extinction will ensue. This happens as the difference between the mean of the population and the optimum value increases.

Lynch & Lande (1993) used the model to consider the magnitude of environmental changes to which populations might adapt, and considered the effects of population size (Figure 7.3) in this process. They suggested that, if genetic variance in a quantitative trait is determined solely by an input from mutation and populations are relatively large, populations may only be able to adapt to a 1% change in the environment per generation. This would not allow populations to adapt to rapidly changing environments. In particular, long-lived species would seem to have little chance of countering the predicted effects of global climatic changes, because these species are likely to encounter environmental changes greater than 1% in their lifetime.

However, we can only use such predictions if the model is realistic. Some of the important variables in Figure 7.3 are difficult to determine, particularly the effects of dispersal and species interactions. In addition, the model makes assumptions about the genetic basis of variation underlying a trait. For instance, Lynch & Lande (1993) assumed that genotypic variance is controlled by numerous genes with small effects. As we have discussed in Chapters 3 and 6, there is insufficient evidence to decide between genetic models of adaptation based on a few genes or numerous genes. Particularly when selection pressures are high, large phenotypic changes often take place on the basis of only a few genes. This is illustrated by the ability of pest insects to rapidly evolve resistance to agricultural chemicals. Theoretical models are therefore useful in highlighting relevant evolutionary parameters, but they have only a limited use for predictive purposes. Any population with a negative growth rate and without much genetic variance is in danger of extinction, but how much genotypic variance does a population need?

Even a large genotypic variance may not guarantee that a population avoids extinction because non-genetic factors can become overriding in importance. This view has been emphasized by researchers working on species with restricted distributions. When populations decline to small numbers, they will often be unable to recover for non-genetic reasons. The density of a population may be so low that individuals have difficulty finding a mate. Low densities may also prevent individuals coming together to fend off predators or competitors. Populations can therefore decline even when they are genetically variable. In addition, small populations may become extinct because of stochastic factors, particularly those related to environmental fluctuations. Because of short-term fluctuations in the weather or other environmental factors, there are times when most individuals do not reproduce or when there is a high level of mortality in susceptible life-cycle stages. This can be catastrophic for small populations, particularly in the case of species with generations that do not overlap. For instance, Ehrlich (1983) discussed several cases where butterfly populations have become extinct as a consequence of environmental variability, despite the presence of high levels of genetic variation in populations.

It has also been pointed out repeatedly that a low level of genetic variation does not necessarily mean that a population is doomed to extinction because of inbreeding or an inability to counter changing conditions. Many laboratory strains of *Drosophila* and mice are viable and vigorous, at least in a laboratory environment, even though they are largely homozygous. If a population lacks genetic variability, its size and density may still be determined by non-genetic factors rather than a lack of genetic variation. For instance, cheetahs have low levels of genetic variation, and this was thought to be associated with a high level of disease susceptibility and a high mortality rate in natural and captive populations. However, Caro & Laurenson (1994) found that a high mortality rate of cheetah cubs in a natural population is mainly due to predation, and to a lesser extent to cubs being abandoned by mothers and environmental catastrophes. In contrast, few cubs appear to be genetically inviable as a consequence of inbreeding or genetic homozygosity. In the short-term, demographic factors may therefore be more likely to contribute to the extinction of cheetahs than homozygosity *per se.*

Despite these arguments, genetic variation must be the *ultimate* factor limiting the persistance of populations under changing environments. Even though genetic homozygosity may not contribute directly to mortality in cheetahs and other species, genetic variation will still determine the ability of these populations to adapt to changing conditions. If cheetahs had been more genetically variable, it is possible that genes countering the effects of predation or stressful conditions could have been selected, leading to a decrease in the levels of mortality currently seen in cheetah populations. Even though ecological factors can provide a proximate explanation for high mortality levels in populations, a lack of genotypic variance still represents the ultimate cause. We have previously made the same point with respect to species borders in Chapter 4.

Because the persistence of populations may ultimately depend on genetic variation, it remains important to maximise the potential for populations to adapt at least partially to changing conditions. This will depend on the persistence of genes important in countering environmental stresses. We need to consider two issues: how to minimise the loss of genetic variation that can occur because of habitat fragmentation, and how to maintain the appropriate type of variation in a population so that it can evolve in response to future changes.

Habitat fragmentation and evolutionary change

As human populations increase in size, we encroach increasingly on natural habitats. Our activities destroy many habitats and severely curtail the areas available for organisms. A consequence of this is the increasing fragmentation of natural populations. As less area becomes available, stretches of forests, grasslands or other ecosystems become restricted to small pockets of land that are not suitable for farming and are therefore the most likely to be set aside as reserves.

These fragmented areas can become like islands surrounded by unfavourable habitats. As they are no longer buffered by the surrounding environment, the physical environment inside fragments can also change. Groom & Schumaker (1993) considered the case of old-growth temperate rainforest in the Olympic Peninsula of Washington State, USA. They found changes in several abiotic variables associated with fragmentation. The extent to which these changes have penetrated into fragments of old-growth forest ranged from several hundred metres for wind speed and around 200 metres for humidity, to less than 100 m for soil temperature, moisture and short-wave radiation. Environmental changes at the edges of fragments are known as 'edge effects', and they influence the distribution of species within fragments. The greater desiccation stress at edges restricts sensitive species such as carrion and dung beetles to forest interiors, while high light levels at edges favour pioneer species that outcompete species of trees predominant in the interior of fragments.

Edge effects are also known in Amazonian rainforests (Lovejoy *et al.*, 1986). In the early morning, relative humidity at edges of reserves can be 5% lower than 100m inside reserves, and this difference can increase to 20% at midday. Air temperatures are up to 5°C higher at edges at this time. Such substantial differences in humidity/temperature gradients from the edge to the centre of reserves can affect entire reserves ≤10 ha in area. These edge effects have a range of biological consequences (Table 7.1) with immediate effects on plants and animals, followed by second and third order changes following biotic interactions. In addition, habitat fragments tend to become progressively more degraded because edges are invaded by undesirable exotic species such as weeds that flourish in the surrounding areas (Quinn & Karr, 1993).

From an evolutionary perspective, habitat fragmentation enhanced by edge effects can be important because this process influences genetic variation. As fragmentation occurs, populations become reduced in size and relatively isolated from other populations. The decrease in size is likely to reduce genetic variability within populations, particularly when populations fall below 100 individuals and the effect of genetic drift becomes marked. In addition, inbreeding effects may become evident when fragments remain isolated, since an increased proportion of matings necessarily occurs between related individuals. Several studies now exist indicating inbreeding effects in natural populations. Keller *et al.* (1994) examined inbreeding levels in an isolated population of song sparrows, *Melospiza melodia*, that underwent two sharp declines in population size following severe winters. They found that inbreeding levels decreased after both population bottlenecks, indicating that inbred individuals were less likely to survive the stressful periods. Madsen *et al.* (1996) considered an isolated population of the adder *Vipera berus*, where heterozygosity levels were low and individuals were related as a consequence of the small population size. Inbreeding effects were apparent from a small litter size and a relatively high proportion of inviable offspring when compared to populations from other areas. Moreover, the introduction of males from other areas led to a marked increase in the production of viable offspring, consistent with the hypothesis that the expression of deleterious genes in the isolated population was responsible for inviability. There is also

Table 7.1. *A list of changes that can occur at edges of forest reserves*

Changes are divided into three classes, depending on whether they represent physical changes, biological changes occuring directly in response to the physical changes, or more indirect biological changes.

Class	Description of change
Changes in the physical environment	Changes in mean temperature and increased temperature fluctuations
	Decreased relative humidity at margins and increased fluctuations
	Increased penetration of light
	Increased exposure to wind
Direct biological changes	Elevated tree mortality leading to standing dead trees
	Falling trees on windward margin
	Reduced leaf fall at margins
	Flourishing plant growth at margins
	Reduced bird populations at margins
	Crowding in refugee animal populations
Higher order biological changes	Increased populations of some insects (e.g., butterflies adapted to high light intensities), disturbance of populations of other insects in forest interior
	Enhanced survival of insectivorous species

Source: Modified from Lovejoy *et al.*, 1986.

evidence for inbreeding depression in small plant populations (eg., Heschel & Paige, 1995).

It is difficult to predict the overall effects of a small population size on the ability of populations to adapt to environmental changes. As noted above, decreases in genetic variability due to genetic drift will reduce the ability of individual populations to respond to environmental changes. However, a decrease in variability within a fragment may be countered by migration between fragments. If N is the size of a population and m is the proportion of this population that consists of migrants, then it can be shown that only limited genetic differentiation will develop between populations if Nm is greater than 3–5. In other words, populations will not diverge much via genetic drift providing they exchange at least a few migrants every generation. This means that even a low level of migration can be effective in preventing divergence between populations, particularly when N is large. We should emphasize that N here refers to the 'effective' population size rather than its actual size. The effective size of a population only includes individuals that contribute genes to the next generation and can be much lower than the census size. If migration prevents much divergence among populations (Nm >3–5), fragmentation may decrease overall levels of genetic variation, as there is a reduction in the total number of individuals. However, if migration levels are low then genetic divergence among populations is expected to occur.

Conversely, there are two ways in which fragmentation may enhance the ability of a species to evolve in response to an environmental change. One possibility is that fragmentation increases overall levels of genetic variation in a species. We can illustrate this with a situation where a species consists of a single population or a series of small populations (known as 'demes') that are partially isolated. In the former case, genetic drift can lead to changes in the frequencies of alleles. Some alleles may eventually be lost from the population or become extremely rare, reducing overall levels of genetic variation. In the latter case, changes in the frequency of alleles as a consequence of drift will be more rapid within a deme because individual demes are smaller than the single population. However, the same alleles will not necessarily be lost in all demes because changes due to genetic drift are random. This can lead to large genetic differences developing between demes, and a high level of variation overall. As a consequence, more allelic variability may be maintained in species consisting of a series of fragmented populations compared to species consisting of one large population.

Holsinger (1993) suggested that some degree of fragmentation can be effective in maintaining genetic variation when traits are under stabilizing selection and affected by several genes. As we have mentioned previously, the intermediate phenotype has the highest fitness under this form of selection (Figure 3.9). When several genes affect a trait, the same intermediate phenotype can be produced by many combinations of alleles. As demes diverge and evolve independently of each other, different combinations of genes producing the same optimal phenotype will be favoured in different populations. In contrast, the same combination of genes will be favoured if there is no fragmentation, reducing the overall genetic diversity in the unfragmented population. As long as there is some gene exchange between demes, alleles favoured in different demes can eventually be available for adaptive responses to environmental changes. Fragmentation could therefore increase rates of evolutionary change.

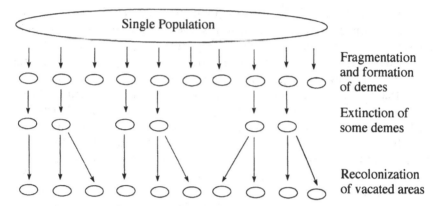

Figure 7.4. Outline of the process of interdemic selection.
Fragmentation of a single population results in a series of
subpopulations (demes). Extinction of some demes can result in
recolonization by individuals from other demes that are better adapted
to an environmental change.

The second way fragmentation can facilitate evolution is via a process
known as 'interdemic selection'. If demes differ genetically, some are more
likely to adapt to an environmental change than others as illustrated in Figure
7.4. This can lead to extinction of demes in some fragments, followed by recolonization of these fragments by individuals from a surviving deme that has
successfully adapted. Once populations have become re-established in all fragments, they will, on average, be better adapted to the new environment.
Genetic divergence among fragments can therefore form the basis of an adaptive process.

That interdemic selection can cause evolutionary change was first suggested
by Sewall Wright (1931) in his 'shifting balance' theory of evolution. Although
the application of this theory to natural populations remains controversial,
there is some evidence that interdemic selection facilitates evolutionary
change. We can illustrate this using an experiment carried out by Goodnight
(1985). He set up demes of the cress, *Arabidopsis thaliana*, and compared the
efficacy of interdemic selection and selection among individuals within demes
in changing the leaf area of this species. For populations under interdemic
selection, only seed from demes with the highest or lowest leaf areas were used
for the next generation. Other demes were assumed to have gone 'extinct'. This
process was effective in increasing and decreasing leaf area (Figure 7.5). In
contrast, selection among individuals within demes did not result in consistent changes in leaf area, and there was not much divergence between lines
selected in opposing directions. Morphological changes were therefore
achieved more readily when selection acted on genetic differences between
demes rather than on variation within demes.

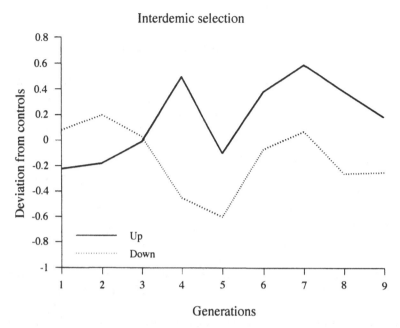

Interdemic selection

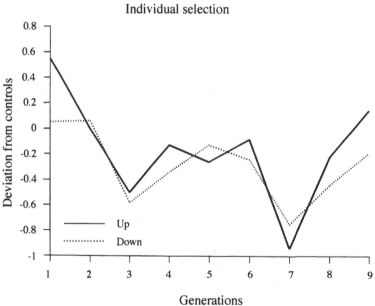

Individual selection

Figure 7.5. A comparison of the response of the cress, *Arabidopsis thaliana*, to group selection and individual selection for increased and decreased leaf area. Although the response to selection fluctuates, there is a clear separation of lines selected up and down when interdemic selection was practised, but not when selection was undertaken among individuals (From Goodnight, 1985).

Nevertheless, it is not known if this type of process is likely in nature. Whether or not responses to selection are increased because of fragmentation depend critically on rates of migration among demes, population sizes and rates of deme extinction. If extinction rates are high and migration rates are low, the distribution of a species may eventually be reduced because vacant fragments are not recolonized. As the overall number of demes becomes reduced, levels of genetic variation within a species will fall and the species' ability to respond to changing conditions may be curtailed. On the other hand, if migration levels are high, genetic differences among fragments are unlikely to be large enough to cause interdemic selection. Fragmentation may therefore only increase evolution when the population size of fragments is relatively small and when there is some migration.

There is evidence that species may have trouble colonizing suitable sites. For instance, the butterfly, *Euphydryas gilletii*, has a restricted distribution in montane meadows of parts of North America, but its host plant occurs much more widely. To determine if dispersal limited colonization, butterflies were transplanted by Holdren & Ehrlich (1981) from a montane site in Wyoming to other montane sites a few hundred kilometres away. The population at one of these sites became established and expanded, indicating that the habitat was suitable and the absence of this species was related to its limited dispersal ability. There are many other examples of dispersal ability limiting range expansion.

Even when individuals reach vacant fragments, they may be unable to colonize it because of the presence of a competing species. There are numerous examples where the removal of one species allows the expansion of a competing species at the same site (Schoener, 1983). When this happens, reinvasion by a competitor is likely to be difficult and slow.

The effects of non-genetic factors on population size will also reduce the efficacy of interdemic selection. The size of populations is closely related to their risk of extinction by the types of environmental fluctuations we discussed above. This is clearly evident in studies of island populations of several animals (Pimm, 1991). For instance, large populations of birds on islands off the coast of Britain and Ireland are much less prone to extinction than small populations (Figure 7.6). In addition, the variability in a population's size affects the risk of a population becoming extinct. Populations that show large fluctuations in size are expected to be more prone to extinction than populations with a more constant size. When environmental changes occur, stress levels will not be uniform across all areas occupied by a fragmented species. Some demes will become smaller in size or experience greater fluctuations in size than others, irrespective of their genetic make-up. This can lead to chance extinction of populations even when they contain highly-adapted individuals. These types of factors may counter any increase in the rate of adaptation stemming from interdemic selection.

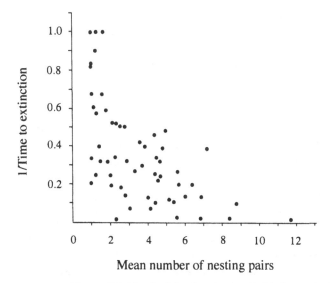

Figure 7.6. Yearly risk of extinction in bird populations from small islands off the coast of Britain and Ireland. The risk of extinction is expressed as the reciprocal of the time that a bird population persisted on an island (i.e., large values represent an increased risk), and this value is plotted against the population size expressed as the mean number of nesting pairs. (From Pimm *et al.*, 1988).

In summary, these considerations make it unlikely that evolutionary changes will be increased by fragmentation. In any case, many species already have fragmented distributions, so additional fragmentation is unlikely to increase levels of genetic divergence among demes (Holsinger, 1993).

Units for conservation

Conservation strategies are aimed at minimising the loss of biodiversity. Much of the recent emphasis in conservation genetics has been on the identification of species and populations that should be a priority for conservation to maintain biodiversity. In particular, it has been proposed that information from DNA sequences could be used to help to conserve the genetic heritage within a species by identifying which populations have evolved independently of each other. These populations are more likely to possess different genetic information.

The approach is outlined by Avise (1994) and Moritz (1994). Phylogenies are constructed from DNA data to show the extent to which populations have genetically diverged. Those populations that are the most distinct genetically are considered to have a high priority. In particular, if populations carry different alleles in DNA sequences from mitochondrial DNA as well as being different for genes encoded by nuclear DNA, they are considered to have the highest conservation value.

As the environment changes, will this strategy maximise the likelihood that a species persists by undergoing evolutionary changes? Is one of the conserved populations fixed for alternate alleles the most likely to evolve? We suspect that this will not always be true. By only conserving the most distinct populations as defined at the molecular level, there is an emphasis on population divergence due to genetic drift rather than selection. DNA sequences used in constructing phylogenies are normally assumed to be neutral, meaning that variation in these sequences is unlikely to influence the phenotype of an organism. A population can carry unique alleles solely because it has been kept at a small size for many generations and therefore been exposed to substantial drift. This has two implications. Firstly, a population exposed to drift for a long time may contain very little genetic variation, reducing its ability to evolve under environmental change. Secondly, the likelihood of populations differing in their evolutionary response may be unrelated to divergence at the DNA level. It is not clear if differences in adaptive responses can be predicted by divergence at neutral gene sequences.

There are alternative approaches for assigning conservation priorities on the basis of the ecological or geographic distinctness of the populations. One possibility is to use ecological distinctness, by considering the types of ecological factors to which populations are exposed. These might include the climates in which species occur, or the types of communities with which they are associated. This measure will often be correlated with the degree to which populations have diverged adaptively as they respond to different ecological conditions, although gene flow among populations could be high enough to prevent adaptive changes. Another possibility is to use geographic distinctness, based on whether populations come from areas that are centrally or marginally located within a species' distribution. Adaptive responses to extreme conditions are more likely in marginal populations.

Genetic and non-genetic measures of distinctness will often lead to the same conclusions. Populations from different environments that have been geographically isolated for some time are likely to have diverged genetically. However, there are scenarios where the measures lead to different conservation priorities. For instance, consider four populations that are isolated geographically. Two of these are small relict populations occupying habitats similar to that occupied by a much larger third population. A fourth population is also large but exposed to a different environment. If only two reserves can be established, which populations do we choose? On the basis of molecular data, the two relict populations would probably be selected, whereas the populations from different habitats will probably show the most rapid and divergent evolutionary responses. It therefore seems prudent to consider all three measures when determining the conservation value of populations.

Captive populations: the problem of adaptation

There is an increasing number of species whose existence in the wild is threatened, and whose long-term survival will depend on captive populations held away from their natural habitats. If the aim of captive breeding progams is the eventual release of captive-bred animals into the wild, released populations will need to counter the effects of environmental changes if they are to persist. We need to consider two problems when attempting to maximise this potential. First, captive populations are always likely to be small. This can result in inbreeding and the loss of genetic variability. Second, captive populations can undergo evolutionary changes in response to the conditions they experience in captivity. These changes can decrease their performance under field conditions.

To minimise genetic changes in captive environments, we need to examine the ways in which these environments are likely to affect populations. Captive environments may be optimal for animals in many ways. Animals will often be bred under favourable temperature and humidity conditions, and will normally be fed high-quality food. This means that animals will not be exposed to the same environmental extremes that they are likely to experience in nature. Animals could therefore become increasingly sensitive to a range of stresses as they are bred in captivity, and this would be a disadvantage once animals are exposed again to natural conditions. As we have seen, selection in optimal environments can result in tradeoffs with fitness in stressful conditions (Chapter 4), so genes selected in captive environments will not be favoured under natural conditions. On the other hand, captive conditions can also be unfavourable. For instance, captive animals tend to be bred in crowded conditions where their movements are severely restricted. These environments will often impose severe selection on behavioural characteristics that decrease mating success in uncrowded conditions and are likely to reduce the flight response of animals.

Genetic changes in captive environments have been demonstrated for numerous organisms. However, there is not much information on the effects of these changes on the re-establishment of reintroduced organisms in natural habitats. Perhaps the best data on this issue come from the establishment of insects used as biological control agents which are introduced to control pest animals and plants. Myers & Sabath (1980) examined factors influencing the success of these introductions. They found that genetic variability as measured by electrophoretic variation did not predict whether released insects were likely to become established. Instead, Myers & Sabath (1980) found that the time insects had been held in captivity was a more important factor than genetic variability, and they cite several cases where adaptation to artificial conditions seems to have had detrimental effects. Reducing the time animals have to adapt to a captive environment can therefore be important when reintroductions of animals are planned in natural habitats.

Since rearing under captive conditions often cannot be avoided, measures need to be taken to minimise the degree of adaptation to these conditions. Frankham & Loebel (1992) suggested several ways in which this might occur. (1)   The generation time could be lengthened to decrease the number of generations over which selection can act. This could be achieved by lowering the temperature at which organisms are cultured, or by forcing organisms to enter quiescent stages. However, there is a danger of adaptation to conditions that lengthen the generation time and this strategy may be difficult to implement in practice because of problems in maintaining captive populations even under favourable conditions.
(2)   Genes from wild populations could be continuously introduced into the captive population. Even a small amount of gene flow between populations can minimise genetic divergence, but this approach may not be feasible for endangered species.
(3)   The heritability of traits under selection could be reduced so that adaptation to conditions in captivity occurs more slowly. One way of achieving this is to ensure that all members of a captive population contribute equally to each generation. This reduces the effects of selection acting on genetic differences between families (Allendorf, 1993). In reality, this approach may be difficult to implement because some animals simply do not breed in captivity.
(4)   The intensity of selection for adaptation to the captive environment could be lowered. This can be achieved by reducing differences between natural and captive environments, emphasizing temperature, desiccation, and nutrition. If natural environments cannot be simulated, then it might be possible to reduce the selection intensity in captivity by ensuring that as many individuals as possible survive and breed. However, this procedure could favour deleterious genes that have a low fitness in all environments, including those in nature.

Some of these approaches may also help to decrease levels of inbreeding in captive populations. Reducing the number of generations undergone by a population will decrease inbreeding levels because inbreeding effects are cumulative. Introducing unrelated individuals from natural populations will also lower inbreeding levels by increasing levels of heterozygosity, which will help to mask deleterious genes responsible for inbreeding effects. The effects of inbreeding in captive populations can be difficult to evaluate because they can depend on conditions organisms experience. Miller (1994) has shown that inbreeding effects are increased when organisms are exposed to stressful conditions and intense competition. This is probably a reflection of the enhanced heterosis in extreme environments that we discussed in Chapter 2. Conditions in captive environments are likely to be stressful, resulting in an enhanced effect of inbreeding.

Captive populations: maintaining genetic variation and the 'appropriate' genes

In conservation genetics, the usual aim is to preserve as much genetic variation in a population as possible. By maintaining high levels of genetic variation, the probability that a population can adapt to unpredictable future environmental changes is thought to be maximised. This approach is based on the assumption that genes likely to be important in future adaptation cannot be identified. Hence, overall levels of genetic variation are considered the best way of determining the potential of populations to adapt to future changes.

Genetic variation can be characterized by two measures, the proportion of loci that are polymorphic (P) and the mean level of heterozygosity of an individual (H). When allozyme variation is being considered, P is determined from the proportion of loci that have more than one allele (i.e., are polymorphic), although loci where the common allele is present at a frequency greater than 95% are normally considered to be monomorphic. H is computed as the mean heterozygosity across a sample of loci, including loci that are polymorphic. Although these measures of genetic variation are usually correlated, it is possible that a population with a large number of uncommon alleles (high value of P) will have a lower level of heterozygosity than a population with a few polymorphic loci (low P) but where alleles are at intermediate frequencies.

To maintain genetic variability, we need to keep captive populations at a large effective population size to minimise the effects of genetic drift. Recent studies have suggested that the effective size of a captive population may often be much less than the actual census size. For instance, Briscoe *et al.* (1992) found that the effective size of large *Drosophila* populations was only a few per cent of the census size. They suggested that competition among males could be partly responsible. Because the majority of matings in a population are usually obtained by a few dominant males, the number of males that breed and contribute progeny to the next generation can be much smaller than the total number of males in a population. This problem may be minimised by changing the conditions that animals experience to ensure that more males have a chance to obtain matings, or by reducing fluctuations in population size (Frankham, 1995).

In many captive populations, pedigrees can be managed in order to maintain high levels of genetic variation for reintroductions into the wild. Several approaches have been advocated. One of these involves the use of data from allozymes or other loci to identify parents that produce the most genetically diverse offspring. This approach can ensure that reintroduced populations have high levels of heterozygosity at the allozyme loci. However, allele diversity at other loci can be lost when decisions are only made on the basis of these loci. There can be a reduction in P even when high levels of H are maintained. Because it is only practical to characterize a few polymorphic enzyme loci for a particular population, uncommon alleles at other loci are likely to be lost.

Variation at a sample of allozyme loci may be a poor indicator of potential responses to selection. Evidence for this comes from experiments where conspecific populations have been exposed to the same selection pressure. For example, Hoffmann & Cohan (1987) selected conspecific populations of *Drosophila pseudoobscura* for increased resistance to ethanol fumes. These populations were initiated with the same number of founders and showed only minor allozyme differences associated with substantial divergent responses to selection. Despite this, there were marked differences in the responses of the populations. Some *D. pseudoobscura* populations showed an increase in resistance to acetone as a correlated response, whereas others showed a decreased resistance to this chemical, implying that responses to selection were based on different physiological mechanisms.

Another approach that has been advocated involves selecting individuals for reintroduction on the basis of the founders from which they are derived. When captive populations are established, they are usually started with only a few individuals from the wild. By selecting individuals so that founders contribute equally to the reintroduced population, many of the alleles present in founders will be passed on. This approach helps to maintain high levels of P and H. However, by equalizing founder effects irrespective of the number of progeny they produce, there is a possibility that some uncommon alleles will be lost (Haig, Ballou & Derrickson, 1990). We can illustrate this with a situation where founders are heterozygous for different alleles. If one pair of founders have one offspring, only two of the four alleles present in these parents will be passed on to the next generation. However, if another pair of founders produces several progeny, the chances of all four parental alleles being passed on is much greater. Equalizing the contribution of all four founders will result in the loss of some alleles. An adjustment for the size of families is therefore required if allelic diversity is to be maintained, and several ways of achieving this have been proposed (see Hedrick & Miller, 1992). However, there are practical difficulties in implementing this strategy because not all animals may breed in captivity.

Haig *et al.* (1990) tested the efficacy of these techniques in ensuring genetic diversity in a population of Guam rail (*Rallus owstoni*). A wild population of this bird species originally existed on the island of Guam, but this population has become extinct. It is being introduced into a neighbouring island using individuals derived from captive stocks. Haig *et al.* (1990) set up a simulation model where they assumed that all founders contributing to the rail pedigree have two unique alleles. They found that fairly high levels of *H* are maintained by using allozyme data and taking founders into account. However, they showed that higher levels of allele diversity can be maintained by considering founders and adjusting for their family sizes than by considering allozyme variation.

Whether a high level of *H* or *P* is more important for future adaptation to environmental stresses has not been tested. Evidence for an association between allozyme heterozygosity and fitness is often cited as support for maintaining a high level of *H* (e.g., Soulé & Wilcox, 1980). In particular, there often seems to be an association between an individual's heterozygosity summed over several enzyme loci, and its growth rate, especially under moderately stressful conditions (Chapter 3). However, this association is not detected in all groups of organisms. Moreover, even if such an association existed, it does not follow that heterozygous individuals have a higher fitness than all homozygous individuals (see Hoffmann & Parsons, 1991; Hedrick & Miller, 1992) or that highly heterozygous species are more resistant of stressful conditions than species with lower heterozygosity levels (eg., Kopp *et al.*, 1994). On the other hand, the usefulness of *P* depends on whether adaptation involves genes that are initially present at a low frequency. It is not clear at the moment if genes that are initially rare are important in adaptive responses to environmental changes.

As an alternative approach to the non-directed conservation of genetic diversity, Hughes (1991) has suggested that genetic variation could be preserved at specific loci that are likely to be particularly important for future adaptation. Hughes (1991) focussed on genetic polymorphism at the major histocompatibility (MHC) locus. Variation at this locus is known to be under natural selection because it is associated with resistance to diseases. Hence, the maintenance of variation at the MHC locus could help to ensure survival of populations faced with new pathogens. A similar process could be envisaged for genes important in responding to changes in environmental conditions, such as those controlling the production of heat shock proteins, although specific loci are probably more difficult to identify than in the case of disease resistance.

This approach has been criticised by others. Vrijenhoek & Leberg (1991) pointed out that there are many loci other than MHC involved with disease resistance. Moreover, traits other than disease resistance will also be under selection. By considering only one locus, genetic variation at other loci is not guaranteed, and overall levels of genetic diversity will therefore decrease. This approach suffers from the same problem as the one discussed above that focusses on a few polymorphic allozyme loci.

Instead of focussing on specific genotypes, it is possible to focus on phenotypic variation. By considering phenotypic variation rather than specific loci, favourable genes at several loci are likely to be selected, thereby partly overcoming a limitation of the approach advocated by Hughes (1991). This approach might be useful if we could be certain that a particular environmental stress is of overriding importance in the successful establishment of a population. By analogy, useful strains of predatory mites have been selected for resistance to insecticides, because this environmental stress determines the successful establishment of mite populations in environments where insecticides are applied (Hoy, 1985).

Selection for specific phenotypes may be feasible when conditions in a reserve have deteriorated because of specific stresses arising from human activities. For instance, endangered plants might be introduced into environments where soils are contaminated by heavy metals. We may find it more difficult to identify specific phenotypes adapted to future climatic changes, although there are some candidate traits. For instance, in beef cattle it is possible to select for resistance to tropical conditions by selecting for a reduction in basal metabolic rate (Frisch, 1981). However, we should emphasize that a directed approach will often be unproductive because the key environmental stresses populations are likely to encounter cannot be identified. Directed approaches should therefore be used cautiously (Miller, 1994).

Two other strategies that are partially directed may be useful in the management of captive populations to be reintroduced into nature. First, useful genes for future adaptation to climatic changes can be conserved by obtaining founders for captive populations from populations at ecological or geographic margins (Hoffmann & Parsons 1991; Lesica & Allendorf 1995; Parsons, 1995). Marginal populations are likely to have been selected for environmental stress resistance. The possibility of generalized stress resistance associated with factors such as a reduction in metabolic rate (Chapter 5) means that genotypes selected under extremes of one type of environmental stress could show increased resistance to other stresses. In contrast, genes increasing stress resistance are likely to be at lower frequencies in more favourable environments than at species margins (Chapter 4). This strategy will not necessarily conflict with attempts to maintain overall levels of genetic variation because levels do not normally differ between central and marginal populations (Chapter 4).

Second, it may be possible to remove individuals that have a particularly low fitness during the maintenance of captive populations as suggested by Frankham, Yoo & Sheldon (1988). This approach was successfully used by these workers to select for increased ethanol resistance in *Drosophila melanogaster*. When selection is carried out on a trait in one direction, the fitness of selected lines often decreases. This is because some genes have large deleterious effects on fitness in most environments. These genes can increase in selected lines because they also have large effects on the selected trait. To counter this, Frankham *et al.* (1988) discarded individuals with a low fitness each generation. They found that lines selected using this procedure still showed a high level of ethanol resistance but did not suffer a decrease in fitness. A culling process in managed or captive populations could ensure that genes with large deleterious effects in a majority of environments do not become common in a population. This process may be undertaken at the same time as inbreeding to increase the chances of deleterious genes becoming expressed. For instance Backus *et al.* (1995) showed that a combination of inbreeding and subsequent selection led to stocks of houseflies with a relatively high level of fitness.

A final point concerns the nature of environments encompassed by reserves. If genotypes that are useful in countering stressful conditions persist in marginal habitats, then it may be beneficial to locate reserves in areas that include these habitats. This approach could have an additional benefit in that high levels of genetic diversity tend to persist longer in heterogeneous environments. Hedrick (1986) reviewed a number of experiments comparing the effects of heterogeneous environmental conditions versus homogeneous conditions on the maintenance of genetic variation. In general, higher levels of variation tended to persist as more aspects of the environment were varied. More recent data associating levels of genetic variation with environmental marginality and environmental variability have been obtained by Nevo and coworkers (Nevo *et al.* 1995).

We conclude this section on a negative note by emphasizing that these considerations may only be useful in the conservation of some species. Genetic strategies will have little impact on the survival of a species if the entire ecosystem where the species occurs is threatened by pollution or global warming. Many species cannot be conserved in captive populations because it will not be possible to breed them in captivity. At least among invertebrates, species that can be bred will be among the more stress-resistant species, and therefore more likely to persist in the long run.

Summary

Rapid increases in global temperature and precipitation have been predicted. Climatic changes are likely to vary considerably between regions. Changes will be exacerbated by the effects of human activities including deforestation and pollution.

Increases in temperature may directly cause mortality in some species. However, many effects will be indirect, such as climatic effects on timing of reproduction and on the incidence of parasites and diseases. The effects on organisms will be increased by multiple stresses and interactions between these stresses.

Changes in the distribution and abundance of stress-sensitive organisms including amphibia, fish and lichens can be used to detect environmental effects. Phenotypic changes within populations of organisms can also be used to monitor environmental effects. Stress-sensitive genotypes can be useful in monitoring the effects of specific pollutants in the field or laboratory, but a range of strains needs to be considered to make realistic predictions for a species.

Genetic models can predict likely adaptive responses to climatic changes. These suggest that successful adaptation will be unlikely if environments change rapidly. However, models so far developed do not allow for rare genes with major effects that may underlie adaptative responses to some environmental changes. Demographic factors can override genetic factors in determining the persistence of populations.

The fragmentation of habitats is associated with edge effects that lead to further deterioration of habitats since abiotic stresses are increased. The fragmentation of habitats could, in theory, increase rates of evolutionary responses to environmental changes, because of the persistence of higher overall levels of variation and because of selection among demes from different fragments. However, fragmentation will often decrease the likelihood of adaptation as species become split into small isolated populations that exist in deteriorating habitats.

Species maintained in captive populations are likely to undergo genetic adaptation to captive environments which tend to be more benign than those in nature. Both this process and the loss of genetic diversity in captivity can, in theory, be minimised. Monitoring allozyme variation is a less useful way of maintaining diversity than equalizing the genetic contribution of founders.

To maximise the potential for future adaptation to environmental changes, it may be more important to maintain diverse alleles than high levels of heterozygosity. Focussing on the conservation of variation at specific loci is probably not a particularly useful strategy. However, it may be appropriate in some instances to select specific phenotypes when particular environmental stresses are of overriding importance. Removing individuals carrying deleterious genes, and ensuring that managed populations include stress-resistant genotypes from marginal environments may be useful strategies for ensuring evolution in response to future environmental changes.

Decisions about which populations and species to conserve to maintain biodiversity need to take environmental change into account. Conservation priorities should be decided on ecological as well as genetic information. An emphasis on stress-resistant populations could be useful when attempting to conserve biodiversity in the face of environmental change.

References

Allen, B.D. & Anderson, R.Y. (1993). Evidence from western North America for rapid shifts in climate during the last glacial maximum. *Science*, **260**, 1920–3.

Allendorf, F. (1993). Delay of adaptation to captive breeding by equalizing family size. *Conservation Biology* **7**, 416–9.

Amundsen, T. & Stokland, J.N. (1988). Adaptive significance of asynchronous hatching in the shag: a test of the brood reduction hypothesis. *Journal of Animal Ecology*, **57**, 329–44.

Andersen, N.M. (1993). The evolution of wing polymorphism in water striders (Gerridae): a phylogenetic approach. *Oikos*, **67**, 433–43.

Andersson, S & Shaw, R.G. (1994). Phenotypic plasticity in *Crespis tectorum* (Asteraceae): genetic correlations across light regimes. *Heredity*, **72**, 113–25.

Andrewartha, H.G. & Birch, L.C. (1954). *The Distribution and Abundance of Animals*. Chicago: University of Chicago Press.

Arking, R., Buck, S., Wells, R.A. & Pretzalff, R. (1988). Metabolic rates in genetically based long-lived strains of *Drosophila*. *Experimental Gerontology*, **23**, 59–76.

Arthur, W. (1984). *Mechanisms of Morphological Evolution*. Chichester: John Wiley & Sons.

Arthur, W. (1987). *The Niche in Competition and Evolution*. New York: John Wiley & Sons.

Asami, T. (1993). Interspecific differences in desiccation tolerance of juvenile land snails. *Functional Ecology*, **7**, 571–7.

Atkinson, D. (1994). Temperature and organism size – a biological law for ectotherms. *Advances in Ecological Research*, **25**, 1–58.

Audo, M.C. & Diehl, W. (1995). Effect of quantity and quality of environmental stress on multilocus heterozygosity-growth relationships in *Eisenia fetida* (Annelida: Oligochaeta). *Heredity*, **75**, 98–105.

Avise, J.C. (1994). *Molecular Markers, Natural History and Evolution*. Chapman & Hall.

Backus, V.L., Bryant, E.H., Hughes, C.R. & Meffert, L.M. (1995). Effect of migration or inbreeding followed by selection on low-founder-number populations: implications for captive breeding programs. *Conservation Biology*, **9**, 1216–24.

Bantock, C.R. & Price, D.J. (1975). Marginal populations of *Cepaea memoralis* (L.) on the Brendon Hills, England. I. Ecology and ecogenetics. *Evolution*, **29**, 267–77.

Barnett, S.A. & Coleman, E.M. (1960). Heterosis in F1 mice in a cold environment. *Genetical Research (Camb.)*, **1**, 25–38.

Barton, N.H. (1989). Founder effect speciation. In *Speciation and its Consequences*, ed. D. Otte & J.A. Endler, pp. 229–56. Sunderland, Massachusetts: Sinauer Associates.

Bateman, K.G. (1959). The genetic assimilation of four venation phenocopies. *Journal of Genetics*, **56**, 443–7.

Bateman, M.A. (1967). Adaptations to temperature in geographic races of the Queensland fruit fly, *Dacus (Strumeta) tryoni*. *Australian Journal of Zoology*, **15**, 1141–61.

Bennington, C.C. & McGraw, J.B. (1995). Natural selection and ecotypic differentiation in *Impatiens pallida*. *Ecological Monographs*, **65**, 303–23.

Benton, M.J. (1987). Progress and competition in macroevolution. *Biological Reviews*, **62**, 305–38.

Benton, T.G. & Grant, A. (1996). How to keep fit in the real world: elasticity analyses and selection pressures on life histories in a variable environment. *American Naturalist*, **147**, 115–39.

Benton, T.G., Grant, A. & Clutton-Brock, T.H. (1995). Does environmental stochasticity matter? Analysis of red deer life-histories on Rum. *Evolutionary Ecology*, **9**, 559–74.

Berrigan, D. & Koella, J.C. (1994). The evolution of reaction norms: simple models for age and size at maturity. *Journal of Evolutionary Biology* **7**, 549–66.

Berrigan, D. & Charnov, E.L. (1994). Reaction norms for age and size at maturity in response to temperature: a puzzle for life historians. *Oikos*, **70**, 474–8.

Billington, H.L. & Pelham, J. (1991). Genetic variation in the date of budburst in Scottish birch populations: implications for climatic change. *Functional Ecology*, **5**, 403–9.

Blaustein, A.R., Hoffman, P.D., Hokit, D.G., Kiesecker, J.M., Walls, S.C. & Hays, J.B. (1994). UV repair and resistance to solar UV-B in amphibian eggs: A link to population declines? *Proceedings of the National Academy of Sciences USA*, **91**, 1791–5.

Blaustein, A.R. & Wake, D.B. (1990). Declining amphibian populations: a global phenomenon? *Trends in Ecology and Evolution*, **5**, 203–4.

Blaxter, K. (1989). *Energy Metabolism in Animals*. New York: Cambridge University Press.

Blows, M.W. & Hoffmann, A.A. (1993). The genetics of central and marginal populations of *Drosophila serrata*. I. Genetic variation for stress resistance and species borders. *Evolution*, **47**, 1255–70.

Blows, M.B. & Hoffmann, A.A. (1996). Evidence for an association between non-additive genetic variation and extreme expression of a trait. *American Naturalist* **148**, 576–87.

Borodin, P.M. (1987). Stress and genetic variability. *Genetika*, **23**, 1003–10.

Boyce, M.S. & Perrins, C.M. (1987). Optimizing great tit clutch size in a fluctuating environment. *Ecology*, **68**, 142–53.

Bradley, B.P. (1978). Genetic and physiological adaptation of the copepod *Eurytemora affinis* to seasonal temperature. *Genetics*, **90**, 193–205.

Brakefield, P.M. (1987). Industrial melanism: do we have the answers? *Trends in Ecology and Evolution*, **2**, 117–22.

Brakefield, P.M. & Lees, D.R. (1987). Melanism in *Adalia* ladybirds and declining air pollution in Birmingham. *Heredity*, **59**, 273–7.

Brasier, M.D. (1991). Nutrient flux and the evolutionary explosion across the Precambrian – Cambrian boundary interval. *Historical Biology*, 5, 85–93.

Briggs, J.C. (1991). A Cretaceous–Tertiary mass extinction? *BioScience*, 41, 619–24.

Briscoe, D.A., Malpica, J.M., Robertson, A., Smith, G.J., Frankham, R., Banks, R.G. & Barker, J.S.F. (1992). Rapid loss of genetic variation in large captive populations of *Drosophila* flies: implications for the genetic management of captive populations. *Conservation Biology*, 6, 416–25.

Brodie, E.D., Moore, A.J. & Janzen, F.J. (1995). Visualizing and quantifying natural selection. *Trends in Ecology and Evolution*, 10, 313–8.

Brown, A.H.D., Marshall, D.R. & Munday, J. (1976). Adaptedness of variants at an alcohol dehydrogenase locus in *Bromus mollis* L. (soft bromegrass). *Australian Journal of Biological Sciences*, 29, 389–96.

Brussard. P.F. (1984). Geographic patterns and environmental gadients: the central – marginal model in *Drosophila* revisited. *Annual Review of Ecology and Systematics*, 15, 25–64.

Bulmer, M.G. (1985). Selection for iteroparity in a variable environment. *American Naturalist*, 126, 63–71.

Calow, P. (1992). The three Rs of ecotoxicology. *Functional Ecology*, 6, 617–9.

Caro, T.M. & Laurenson, M.K. (1994). Ecological and genetic factors in conservation: a cautionary tale. *Science*, 263, 485–6.

Carroll, S.B. (1995). Homeotic genes and the evolution of arthropods and chordates. Nature, 376, 479–85.

Carson, H.L. & Templeton, A.R. (1984). Genetic revolutions in relation to speciation phenomena: the founding of new populations. *Annual Review of Ecology and Systematics*, 15, 97–131.

Castilla, A.M. & Bauwens, D. (1991). Thermal biology, microhabitat selection, and conservation of the insular lizard *Podacris hispanica atrata*. *Oecologia*, 85, 366–74.

Caughley, G., Grigg, G.C. & Smith, L. (1985). The effect of drought on kangaroo populations. *Journal of Wildlife Management*, 49, 679–85.

Caughley, G., Short, J., Grigg, G.C. & Nix, H. (1987). Kangaroos and climate: an analysis of distribution. *Journal of Animal Ecology*, 56, 751–61.

Chan, J.W.Y. & Burton, R.S. (1992). Variation in alcohol dehydrogenase activity and flood tolerance in white clover, *Trifolium repens*. *Evolution*, 46, 721–34.

Chapin, F.S. (1980). The mineral nutrition of wild plants. *Annual Review of Ecology and Systematics*, 11, 233–60.

Chappell, M.A. & Snyder, L.R.G. (1984). Biochemical and physiological correlates of deer mouse alpha-chain hemoglobin polymorphisms. *Proceedings of the National Academy of Sciences USA*, 81, 5484–8.

Charlesworth, B. (1994). *Evolution in Age-structured Populations*. 2nd ed. Cambridge: Cambridge University Press.

Clarke, C.A., Clarke, F.M.M. & Dawkins, H.C. (1990). *Biston betularia* (the peppered moth) in West Kirby, Wirral, 1959–1989: updating the decline in *f. carbonaria*. *Biological Journal of the Linnean Society*, 39, 323–6.

Clarke, G.M. (995). Relationships between developmental stability and fitness – application for conservation biology. *Conservation Biology*, 9, 18–24. |

Cohen, D. (1966). Optimizing reproduction in a randomly varying environment. *Journal of Theoretical Biology*, 12, 119–29.

Cole, L.C. (1954). The population consequences of life history phenomena. *Quarterly Review of Biology,* **29**, 103–37.

Conway Morris, S. (1987). The search for the Precambrian – Cambrian boundary. *American Scientist,* **75**, 157–67.

Cooch, E.G. & Ricklefs, R.E. (1994). Do variable environments significantly influence optimal reproductive effort in birds? *Oikos,* **69**, 447–59.

Cooke, B.D. (1977). Factors limiting the distribution of the wild rabbit in Australia. *Proceedings of the Ecological Society of Australia,* **10**, 113–120.

Coope, G.R. (1979). Late Cenozoic fossil Coleoptera: evolution, biogeography, and ecology. *Annual Review of Ecology and Systematics,* **10**, 247–67.

Coulson, S.C. & Bale, J.S. (1992). Effect of rapid cold hardening on reproduction and survival of offspring in the housefly *Musca domestica. Journal of Insect Physiology,* **38**, 421–4.

Cox, R.M. & Hutchinson, T.C. (1980). Multiple metal tolerances in the grass *Deschampsia cespitosa* (L.) Beauv. from the Sudbury smelting area. *New Phytologist,* **84**, 631–47.

Crespi, B.J. (1990). Measuring the effect of natural selection on phenotypic interaction systems. *American Naturalist,* **135**, 32–47.

Cullis, C.A. (1987). The generation of somatic and heritable variation in response to stress. *American Naturalist,* **130**, S62–S73.

Darwin, C. (1859). *On the Origin of Species by means of Natural Selection.* London: Murray.

Dawson, W.R. (1992). Physiological responses of animals to higher temperatures. In *Global Warming and Biological Diversity,* ed. R.L. Peters & T.E. Lovejoy, pp. 158–70. New Haven: Yale University Press.

Dean, A.M. (1994). Fitness, flux and phantoms in temporally variable environments. *Genetics,* **136**, 1481–95.

Derr. J.A. (1980). The nature of variation in life history characters of *Dysdercus bimaculatus* (Heteroptera: Pyrrhocoridae), a colonizing species. *Evolution,* **34**, 548–57.

DiMichele, W.A. & Aronson, R.B. (1992). The Pennsylvanian–Permian vegetational transition: a terrestrial analogue to the onshore–offshore hypothesis. *Evolution,* **46**, 807–24.

Dingle, H. (1981). Geographic variation and behavioral flexibility in milkweed bug life histories. In *Insect Life History Patterns: Habitat and Geographical Variation,* ed. R.F. Denno & H. Dingle, pp. 57–73. New York: Springer-Verlag.

Dobson, A. & Carper, R. (1992). Global warming and potential changes in host – parasite and disease-vector relationships. In *Global Warming and Biological Diversity,* ed. R.L. Peters & T.E. Lovejoy, pp. 201–17. New Haven: Yale University Press.

Donovan, S.K. (ed.) (1989). *Mass Extinctions: Processes and Evidence.* London: Belhaven Press.

Downes, J.A. & Kavanaugh, D.H. (1988). Origins of the North American insect fauna. Introduction and commentary. *Memoirs of the Entomological Society of Canada,* **44**, 1–11.

Dunham, A.E. (1993). Population responses to environmental change: operative environments, physiologically structured models and population dynamics. In

Biotic Interactions and Global Change ed. P.M. Kareiva, J.G. Kingsolver & R.B. Huey, pp. 95–119. Sunderland, Massachusetts: Sinauer Associates.

Dunson, W.A., Wyman, R.L. & Corbett, E.S. (1992). A symposium on amphibian declines and habitat acidification. *Journal of Herpetology*, **26**, 349–52.

Ehrlich, P.R. (1983). Genetics and the extinction of butterfly populations. In *Genetics and Conservation*, ed. C.M. Schonewald- Cox, S.M. Chambers, B. MacBryde & W.L. Thomas, pp. 152–63. Menlo Park, California: Benjamin/Cummings.

Ehrlich, P.R., Murphy, D.D., Singer, M.C., Sherwood, C.B., White, R.R. and Brown, I.L. (1980). Extinction, reduction, stability and increase: the responses of checkerspot butterfly (*Euphydryas*) populations to the California drought. *Oecologia*, **46**, 101–5.

Eldredge, N. & Gould, S.J. (1972). Punctuated equilibria: an alternative to phyletic gradualism. In *Models in Paleobiology* ed. T.J.M. Schopf, pp. 82–115. San Francisco: W.H. Freeman.

Endler, J.A. (1986). *Natural Selection in the Wild*. Princeton: Princeton University Press.

English-Loeb, G.M. (1990). Plant drought stress and outbreaks of spider mites: a field test. *Ecology*, **71**, 1401–11.

Erwin, D.H., Valentine, J.W. & Sepkoski, J.J. (1987). A comparative study of diversification events: the early Paleozoic versus the Mesozoic. *Evolution*, **41**, 1177–86.

Etter, R.J. (1988). Physiological stress and color polymorphism in the intertidal snail *Nucella lapillus*. *Evolution*, **42**, 660–80.

Falconer, D.S. & Mackay, T.F.C. (1995). *Introduction to Quantitative Genetics*, 4th edn. Harlow: Longman.

Feder, W.A. & Shrier, R. (1990). Combination of U.V.-B and ozone reduces pollen tube growth more than either stress alone. *Environmental and Experimental Botany*, **30**, 451–4.

Fellner, R. & Peskova, V. (1995). Effects of industrial pollutants on ectomycorrhizal relationships in temperate forests. *Canadian Journal of Botany*, **73**, S 1310–15.

Finch, S. & Collier, R.H. (1983). Emergence of flies from overwintering populations of cabbage root fly pupae. *Ecological Entomology*, 8, 29–36.

Fisher, N.S. (1977). On the differential sensitivity of estuarine and open-ocean diatoms to exotic chemical stress. *American Naturalist*, **111**, 871–95.

Fisher, R.A. (1930). *The Genetical Theory of Natural Selection*. Oxford: Clarendon Press.

Forbes, V.E. & Depledge, M.H. (1992). Predicting population response to pollutants: the significance of sex. *Functional Ecology*, **6**, 376–81.

Ford, M.J. (1982). *The Changing Climate: Responses of the Natural Fauna and Flora*. London: George Allen and Unwin.

Frankham, R. (1995). Effective population size adult population size ratios in wildlife – a review. *Genetical Research*, **66**, 95–107.

Frankham, R. & Loebel, D.A. (1992). Modeling problems in conservation genetics using captive *Drosophila* populations: rapid genetic adaptation to captivity. *Zoo Biology*, **11**, 333–42.

Frankham, R., Yoo, B.H. & Sheldon, B.L. (1988). Reproductive fitness and artificial selection in animal breeding: culling on fitness prevents a decline in reproductive fitness in lines of *Drosophila melanogaster* selected for increased inebriation time. *Theoretical and Applied Genetics*, **76**, 909–14.

Frisancho, A.R., Frisancho, H.G., Milotich, M., Brutsaert, T, Albalak, R., Spielvogel, H., Villena, M., Vargas, E. & Soria, R. (1995). Developmental, genetic and environmental components of aerobic capacity at high altitude. *American Journal of Physical Anthropology*, **96**, 431–42.

Frisch, J.E. (1981). Changes occurring in cattle as a consequence of selection for growth rate in a stressful environment. *Journal of Agricultural Science*, **96**, 23–38.

Frumhoff, P.C. & Reeve, H.K. (1994). Using phylogenies to test hypotheses of adaptation: a critique of some current proposals. *Evolution*, **48**, 172–80.

Fry, F.E.J. (1958). Temperature compensation. *Annual Review of Physiology*, **20**, 207–24.

Futuyma, D. (1989). Macroevolutionary consequences of speciation. In *Speciation and its Consequences*, ed. D. Otte & J.A. Endler, pp. 557–78. Sunderland, Massachusetts: Sinauer Associates.

Gibbs, H.L. & Grant, P.R. (1987). Adult survivorship in Darwin's ground finch (*Geospiza*) populations in a variable environment. *Journal of Animal Ecology*, **56**, 797–813.

Gibson, G. & Hogness, D.S. (1996). Effect of polymorphism in the *Drosophila* regulatory gene *Ultrabithorax* on homeotic stability. *Science*, **271**, 200–3.

Gilbert, N. (1980). Comparative dynamics of a single-host aphid. I. The evidence. *Journal of Animal Ecology*, **49**, 351–69.

Gillespie, J. (1973). Polymorphism in random environments. *Theoretical Population Biology*, **4**, 193–5.

Glynn, P.W. (1988). El Niño-Southern oscillation 1982–1983. Nearshore population, community, and ecosystem responses. *Annual Review of Ecology and Systematics*, **19**, 309–45.

Goodman, D. (1984). Risk spreading as an adaptive strategy in iteroparous life histories. *Theoretical Population Biology*, **23**, 1–20.

Goodnight, C.J. (1985). The influence of environmental variation on group and individual selection in a cress. *Evolution*, **39**, 545–58.

Gould, S.J. & Eldredge, N. (1993). Punctuated equilibrium comes of age. *Nature*, **366**, 223–7.

Graham, R.W. (1992). Late Pleistocene faunal changes as a guide to understanding effects of greenhouse warming on the mammalian fauna of North America. In *Global Warming and Biological Diversity* ed. R.L. Peters & T.E. Lovejoy, pp. 76–87. New Haven: Yale University Press.

Grant, P.R. (1986). *Ecology and Evolution of Darwin's Finches*. Princeton: Princeton University Press.

Greenwood, S.R. (1987). The role of insects in tropical forest food webs. *Ambio*, **16**, 267–71.

Grime, J.P. (1979). *Plant Strategies and Vegetation Processes*. John Wiley, Chichester.

Groom, M.J. & Schumaker, N. (1993). Evaluating landscape change: patterns of worldwide deforestation and local fragmentation. In *Biotic Interactions and Global Change*, ed. P.M. Kareiva, J.G. Kingsolver & R.B. Huey, pp. 24–44. Sunderland, Massachusetts: Sinauer Associates.

Haig, S.M., Ballou, J.D. & Derrickson, S.R. (1990). Management options for preserving genetic diversity: reintroduction of Guam rails to the wild. *Conservation Biology*, **4**, 290–300.

Haines, A. (1991). Global warming and health. *British Medical Journal*, **302**, 669–70.

Hall, B.G. (1990). Spontaneous point mutations that occur more often when they are advantageous than when they are neutral. *Genetics*, **126**, 5–16.

Hallam, A. (1990). Biotic and abiotic factors in the evolution of early Mesozoic marine molluscs. In *Causes of Evolution: A Paleontological Perspective*. eds. R.M. Ross & W.D. Allman, pp. 249–69. Chicago: University of Chicago Press.

Hallam, A. (1992). *Phanerozoic Sea-level Changes*. New York: Cambridge University Press.

Hanna, J.M. & Brown, D.E. (1983). Human heat tolerance: an anthropological perspective. *Annual Review of Arthropology*, **12**, 259–84.

Harrison, G.A., Tanner, J.M., Pilbeam, D.R. & Baker, P.T. (1988). *Human Biology*, 3rd ed., Oxford: Oxford University Press.

Hartl, D.L., Dijkhuizen, D.E. & Dean, A.M. (1985). Limits of adaptation: the evolution of selective neutrality. *Genetics*, **111**, 655–74.

Harvey, P.H. & Pagel, M.D. (1991). *The Comparative Method in Evolutionary Biology*. Oxford: Oxford University Press.

Hashimoto, T., Otaka, E., Adachi, J., Mizuta, K. & Hasegawa, M. (1993). The giant panda is closer to a bear, judged by α- and β-hemoglobin sequences. *Journal of Molecular Evolution*, **36**, 282–9.

Hastings, A. & Caswell, H. (1979). Role of environmental variability in the evolution of life history strategies. *Proceedings of the National Academy of Sciences USA*, **76**, 4700–3.

Hedrick, P.W. (1986). Genetic polymorphism in heterogeneous environments: a decade later. *Annual Review of Ecology and Systematics*, **17**, 535–66.

Hedrick, P.W. & Miller, P.S. (1992). Conservation genetics: techniques and fundamentals. *Ecological Applications*, **2**, 30–46.

Hersteinsson, P. & Macdonald, D.W. (1992). Interspecific competition and the geographical distribution of red and arctic foxes, *Vulpes vulpes* and *Alopex lagopus*. *Oikos*, **64**, 505–15.

Heschel, M.S. & Paige, K.N. (1995). Inbreeding depression, environmental stress, and population size variation in scarlet gilia (*Ipomopsis aggregata*). *Conservation Biology*, **9**, 126–33.

Hickey, D.A. & McNeilly, T. (1975). Competition between metal tolerant and normal plant populations: a field experiment on normal soil. *Evolution*, **29**, 458–64.

Hinrichsen, D. (1986). Multiple pollutants and forest decline. *Ambio*, **13**, 258–65.

Hochachka, P.W. & Somero, G.N. (1984). *Biochemical Adaptation*. Princeton: Princeton University Press.

Hochachka, P.W., Stanley, C., Matheson, G.O., McKenzie, D.C., Allen, P.S. and Parkhouse, W.S. (1991). Metabolic and work efficiencies during exercise in Andean natives. *Journal of Applied Physiology*, **70**, 1720–30.

Hoffmann, A.A. & Blows, M.W. (1994). Species borders: ecological and evolutionary perspectives. *Trends in Ecology and Evolution*, **9**, 223–7.

Hoffmann, A.A. & Cohan, F.M. (1987). Genetic divergence under uniform selection. III. Selection for knockdown resistance to ethanol in *Drosophila pseudoobscura* populations and their replicate lines. *Heredity*, **58**, 425–33.

Hoffmann, A.A. & Parsons, P.A. (1989). An integrated approach to environmental stress tolerance and life-history variation. Desiccation tolerance in *Drosophila*. *Biological Journal of the Linnean Society*, **37**, 117–36.

Hoffmann, A.A. & Parsons, P.A. (1991). *Evolutionary Genetics and Environmental Stress*, Oxford: Oxford University Press.

Hoffmann, A.A. & Parsons, P.A. (1993). Selection for adult desiccation resistance in *Drosophila melanogaster*: fitness components, larval resistance and stress correlations. *Biological Journal of the Linnean Society*, **48**, 43–54.

Holdren, C.E. & Ehrlich, P.R. (1981). Long range dispersal in checkerspot butterflies: transplant experiments with *Euphydryas gillettii*. *Oecologia*, **50** 125–9.

Holsinger, K.E. (1993). The evolutionary dynamics of fragmented plant populations. In *Biotic Interactions and Global Change*. ed. P.M. Kareiva, J.G. Kingsolver & R.B. Huey. pp. 198–216. Sunderland, Massachusetts: Sinauer Associates.

Hopper, S.D. (1979). Biogeographical aspects of speciation in the southwest Australian flora. *Annual Review of Ecology and Systematics*, **10**, 399–422.

Houghton, P. (1990). The adaptive significance of Polynesian body form. *Annals of Human Biology*, **17**, 19–32.

Houle, D. (1992). Comparing evolvability and variability of quantitative traits. *Genetics*, **130**, 185–204.

Houston, A.I. & McNamara, J.M. (1992). Phenotypic plasticity as a state dependent life-history decision. *Evolutionary Ecology*, **6**, 243–53.

Howarth, F.G. (1993). High-stress subterranean habitats and evolutionary change in cave-inhabiting arthropods. *American Naturalist*, **140**, S65–S77.

Hoy, M.A. (1985). Recent advances in genetics and genetic improvement of the Phytoseiidae. *Annual Review of Entomology*, **30**, 345–70.

Huether, C.A. (1968). Exposure of natural genetic variability underlying the pentamerous corolla constancy in *Linanthus androsoceus* spp. *androsaceus*. *Genetics*, **60**, 123–46.

Huether, C.A. (1969). Constancy of the pentamerous corolla phenotype in natural populations of *Linanthus*. *Evolution*, **23**, 572–88.

Huey, R.B. & Berrigan, D. (1996). Testing evolutionary hypotheses of acclimation. In *Phenotypic and Evolutionary Adaptation to Temperature*, ed. by I.A. Johnston & A.F. Bennett. Cambridge: Cambridge University Press.

Huey, R.B. & Kingsolver, J.G. (1993). Evolution of resistance to high temperature in ectotherms. *American Naturalist*, **142**, S21–S46.

Huey, R.B. & Stevenson, R.D. (1979). Integrating thermal physiology and ecology of ectotherms: a discussion of approaches. *American Zoologist*, **19**, 357–66.

Hughes, H.L. (1991). MHC polymorphism and the design of captive breeding programs. *Conservation Biology*, **5**, 249–51.

Ihlenfeldt, H-D. (1994). Diversification in an arid world: the Mesembryanthemaceae. *Annual Review of Ecology and Systematics*, **25**, 521–46.

Ivanovici, A.M. & Wiebe, R.J. (1981). Towards a working 'definition' of 'stress': a review and critique. In *Stress Effects on Natural Ecosystems*, ed. by G.W. Barrett & R. Rosenberg, pp. 13–27. New York: John Wiley & Sons.

Jablonka, E. & Lamb, M. (1995). *Epigenetic Inheritance and Evolution*. Oxford: Oxford University Press.

Jablonski, D. (1987). Heritability at the species level: analysis of geographic ranges of Cretaceous mollusks. *Science*, **238**, 360–3.

Jablonski, D. (1991). Extinctions: a paleontological perspective. *Science*, **253**, 754–7.

Jablonski, D. & Bottjer, D.J. (1990). The ecology of evolutionary innovation: the fossil record. In *Evolutionary Adaptations*, ed. by M.H. Nitecki, pp. 253–88. Chicago: University of Chicago Press.

Jeffs, P.S., Holmes. E.C. & Ashburner, M. (1994). The molecular evolution of the alcohol dehydrogenase and alcohol dehydrogenase-related genes in the *Drosophila melanogaster* species subgroup. *Molecular Biology and Evolution*, **11**, 287–304.

Jenkins, N.L. & Hoffmann, A.A. (1994). Genetic and maternal variation for heat resistance in *Drosophila* from the field. *Genetics*, **137**, 783–9.

Johannesson, K., Johannesson, B. & Lundgren, U. (1995). Strong natural selection causes microscale allozyme variation in a marine snail. *Proceedings of the National Academy of Sciences USA*, **92**, 2602–6.

Johnson, P.A., Lenski, R.E. & Hoppensteadt, F.C. (1995). Theoretical analysis of divergence in mean fitness between initially identical populations. *Proceedings of the Royal Society of London B*, **259**, 125–30.

Jones, J.S., Coyne, J.A. & Partridge, L. (1987). Estimation of thermal niche of *Drosophila melanogaster* using a temperature-sensitive mutation. *American Naturalist*, **130**, 83–90.

Jones, J.S., Leith, B.H. & Rawlings, P. (1977). Polymorphism in *Cepaea*: a problem with too many solutions. *Annual Review of Ecology and Systematics*, **8**, 109–43.

Jordan, N. (1991). Multivariate analysis of selection in experimental populations derived from hybridization of two ecotypes of the annual plant *Diodia teres* W. (Rubiaceae). *Evolution*, **45**, 1760–72.

Kacser, H. & Burns, J.A. (1981). The molecular basis of dominance. *Genetics*, **97**, 639–66.

Kaitala, A. (1988). Wing muscle dimorphism: two reproductive pathways of the waterstrider *Gerris thoracicus* in relation to habitat instability. *Oikos*, **53**, 222–8.

Kalisz, S. (1986). Variable selection on the timing of germination in *Collinsia verna* (Scrophulariaceae). *Evolution*, **40**, 479–91.

Kareiva, P.M., Kingsolver, J.G. & Huey, R.B. (1993). (eds.) *Biotic Interactions and Global Change*, Sunderland, Massachusetts: Sinauer Associates.

Katz, R.W. & Brown, B.G. (1992). Extreme events in a changing climate: variability is more important than averages. *Climatic Change*, **21**, 289–302.

Kauffman, S.A. (1993). *The Origins of Order*. New York: Oxford University Press.

Kawecki, T.J. (1994). Accumulation of deleterious mutations and the evolutionary cost of being a generalist. *American Naturalist*, **144**, 833–8.

Kawecki, T.J. (1995). Expression of genetic and environmental variation for life history characters on the usual and novel hosts in *Callosobruchus maculatus* (Coleoptera: Bruchidae). *Heredity*, **75**, 70–6.

Kawecki, T.J. & Stearns, S.C. (1993). The evolution of life histories in spatially heterogeneous environments: optimal reaction norms revisited. *Evolutionary Ecology*, **7**, 155–74.

Keller, L.F., Arcese, P., Smith, J.N.M., Hochachka, W.M. & Stearns, S.C. (1994). Selection against inbred song sparrows during a natural population bottleneck. *Nature*, **372**, 356–7.

Kellogg, D.E. (1975). The role of phyletic change in the evolution of *Pseudocubus vema* (Radiolaria). *Paleobiology*, **1**, 359–70.

Kieser, J.A. (1992). Fluctuating odontometric asymmetry and maternal alcohol consumption. *Annals of Human Biology*, **19**, 513–20.

Kindred, B. (1967). Selection for an invariant character, vibrissae in the house mouse. V. Selection on non-Tabby segregants from Tabby selection lines. *Genetics*, **55**, 365–73.

Kingsolver, J.G. & Schemske, D.W. (1991). Path analysis of selection. *Trends in Ecology and Evolution*, **6**, 276–80.

Kitchell, J.A., Clark, D.L. & Gombos, A.M. (1986). Biological selectivity of extinction: a link between background and mass extinction. *Palaios*, **1**, 504–11.

Kleiber, M. (1961). *The Fire of Life: An Introduction to Animal Energetics*. New York: Wiley & Sons.

Koch, P.L., Zachos, J.C. & Gingerich, P.D. (1992). Correlation between isotope records in marine and continental carbon reservoirs near the Paleocene/Eocene boundary. *Nature*, **358**, 319–22.

Koehn, R.K. & Bayne, B.L. (1989). Towards a physiological and genetical understanding of the energetics of the stress response. *Biological Journal of the Linnean Society*, **37**, 157–71.

Kondrashov, A. S. & Houle, D. (1994). Genotype–environment interaction and the estimation of the genomic mutation rate in *Drosophila melanogaster*. *Proceedings of the Royal Society of London Series B*, **258**, 221–7.

Kopp, R.L., Wissing, T.E. & Guttman, S.I. (1994). Genetic indicators of environmental tolerance among fish populations exposed to acid deposition. *Biochemical Systematics and Ecology*, **22**, 459–75.

Korol, A.B. & Iliardi, K.G. (1994). Increased recombination frequencies resulting from directional selection for geotaxis in *Drosophila*. *Heredity*, **72**, 64–8.

Krebs, R.A. & Loeschcke, V. (1994). Effects of exposure to short-term heat stress on fitness components in *Drosophila melanogaster*. *Journal of Evolutionary Biology*, **7**, 39–49.

La Belle, R.P. & Bradley, B.P. (1982). Selection for temperature tolerance during power plant entrainment of copepods. *Journal of Thermal Biology*, **7**, 39–44.

Lack, D. (1954). *The Natural Regulation of Animal Numbers*. London: Oxford University Press.

Lamey, T.C. & Lamey, C.S. (1994). Hatch asynchrony and bad food years. *American Naturalist*, **143**, 734–8.

Lande, R. & Arnold, S.J. (1983). The measurement of selection on correlated characters. *Evolution*, **37**, 1210–26.

Larson, A. (1989). The relationship between speciation and morphological evolution. In *Speciation and its Consequences*, ed. D. Otte & J.A. Endler, pp. 579–98 Sunderland, Massachusetts: Sinauer Associates.

Lenski, R.E. & Bennett, A.F. (1993). Evolutionary response of *Escherichia coli* to thermal stress. *American Naturalist*, **142**, S47–S64.

Lenski, R.E. & Mittler, J.E. (1993). The directed mutation controversy and Neo-Darwinism. *Science*, **259**, 188–94.

Leroi, A.M., Rose, M.R. & Lauder, G.V. (1994). What does the comparative method reveal about adaptation? *American Naturalist*, **143**, 381–402.

Lesica, P. & Allendorf, F.W. (1995). When are peripheral populations valuable for conservation? *Conservation Biology*, **9**, 753–60.

Levin, S.A. & Cohen, D. (1991). Dispersal in patchy environments: the effects of temporal and spatial structure. *Theoretical Population Biology*, **39**, 63–99.

Levin, S.A., Cohen, D. & Hastings, A. (1984). Dispersal strategies in patchy environments. *Theoretical Population Biology*, **26**, 165–91.

Levinton, J. (1988). *Genetics, Paleontology, and Macroevolution*. New York: Cambridge University Press.

Liou, L.W., Price, T., Boyce, M.S. & Perrins, C.M. (1993). Fluctuating environments and clutch size evolution in great tits. *American Naturalist*, **141**, 507–16.

Lipps, J.H. (1986). Extinction dynamics in pelagic ecosystems. In *Dynamics of Extinction*, ed D.K. Elliott, pp. 87–104. New York: Wiley-Interscience.

Lithgow, G.J., White, T.M., Melov, S. & Johnson, T.E. (1995). Thermotolerance and extended life-span conferred by single-gene mutations and induced by thermal stress. *Proceedings of the National Academy of Sciences USA*, **92**, 7540–4.

Liu, E.H., Sharitz, R.R. & Smith, M.H. (1978). Thermal sensitivities of malate dehydrogenase isozymes in *Typha*. *American Journal of Botany*, **65**, 214–20.

Livshits, G. & Kobyliansky, E. (1991). Fluctuating asymmetry as a possible measure of developmental homeostasis in humans: a review. *Human Biology*, **63**, 441–66.

Lodge, D.M. (1993). Species invasions and deletions: community effects and responses to climate and habitat change. In *Biotic Interactions and Global Change*, ed. P.M. Kareiva, J.G. Kingsolver & R.B. Huey, pp. 367–87. Sunderland, Massachusetts: Sinauer Associates.

Loik, M.E. & Noble, P.S. (1993). Freezing tolerance and water relations of *Opuntia fragilis* from Canada and the United States. *Ecology*, **74**, 1722–32.

Long, L.E., Saylor, L.S. & Soulé, M.E. (1995). A pH/UV-B synergism in amphibians. *Conservation Biology*, **9**, 1301–3.

Lovegrove, B.G. (1986). The metabolism of social subterranean rodents: adaptation to aridity. *Oecologia*, **69**, 551–5.

Lovejoy, T.E., Bierregaard, R.V., Jr., Rylands, A.B., Malcolm, J.R., Quintela, C.E., Harper, L.H., Brown, K.S., Jr., Powell, A.H., Powell, G.V.N., Schubart, H.O.R. & Hays, M.B. (1986). Edge and other effects of isolation on Amazon forest fragments. In *Conservation Biology: the Science of Scarcity and Diversity*, ed. M.E. Soulé. pp. 257–85. Sunderland Massachusetts: Sinauer Associates.

Lovett Doust, L., Lovett Doust, J. & Schmidt, M. (1993). In praise of plants as biomonitors – send in the clones. *Functional Ecology*, **7**, 754–8.

Lynch, M. & Lande, R. (1993). Evolution and extinction in response to environmental change. In *Biotic Interactions and Global Change*, ed. P.M. Kareiva, J.G. Kingsolver & R.B. Huey, pp. 234–50. Sunderland, Massachusetts: Sinauer Associates.

Mabberley, D.J. & Hay, A. (1994). Homoeosis, canalization, decanalization, 'characters' and angiosperm origins. *Edinburgh Journal of Botany*, **51**, 117–26.

MacFarlane, W.V. (1976). Aboriginal palaeophysiology. In *The Origin of the Australians*, ed. R.L. Kirk & A.G. Thorne, pp. 183–94. Canberra: Australian Institute of Aboriginal Studies.

Magrath, R.D. (1989). Hatching asynchrony and reproductive success in the blackbird. *Nature*, **339**, 536–8.

Machado-Allison, C.E. & Craig, G.B. (1972). Geographic variation in resistance to desiccation in *Aedes aegypti* and *A. atropalpus* (Diptera: Culicidae). *Annals of the Entomological Society of America*, **65**, 542–7.

Mackay, T.F.C. (1995). The genetic basis of quantitative variation – numbers of sensory bristles of *Drosophila melanogaster* as a model system. *Trends in Genetics*, **11**, 464–70.

Macnair, M.R. (1991). Why the evolution of resistance to anthropogenic toxins normally involve major gene changes: the limits to natural selection. *Genetica*, **84**, 213–9.

Macnair, M.R., Smith, S.E. & Cumbes, Q.J. (1993). Heritability and distribution of variation in degree of copper tolerance in *Mimulus guttatus* at Copperopolis, California. *Heredity*, **71**, 445–55.

MacNicoll, A.D. (1988). The role of altered vitamin K metabolism in anticoagulant resistance in rodents. In *Current Advances in Vitamin K Research*, ed. J.W. Suttie, pp. 407–17. Amsterdam: Elsevier.

MacPhee, D.G. (1993). Directed evolution reconsidered. *American Scientist*, **81**, 554–61.

MacPhee, D.G. & Ambrose, M. (1996). Spontaneous mutations in bacteria: chance or necessity? *Genetica*, **97**, 87–101.

Madsen, T., Stille, B. & Shine, R. (1996). Inbreeding depression in an isolated population of adders *Vipera berus*. *Biological Conservation*, **75**, 113–8.

Mani, G.S. (1990). Theoretical models of melanism in *Biston betularia* – a review. *Biological Journal of the Linnean Society*, **39**, 355–71.

Manning, W.J. & Feder, W.A. (1980). *Biomonitoring Air Pollutants with Plants*. London: Applied Science.

Markow, T.A. (1995). Evolutionary ecology and developmental instability. *Annual Review of Ecology and Systematics*, **40**, 105–20.

Masters, J.C. & Rayner, R.J. (1993). Competition and macroevolution: the ghost of competition yet to come? *Biological Journal of the Linnean Society*, **49**, 87–98.

Maynard Smith, J. (1982). *Evolution and the Theory of Games*. Cambridge: Cambridge University Press.

McClintock, B. (1984). The significance of responses of the genome to challenge. *Science*, **226**, 792–801.

McDonald, J.F. (1990). Macroevolution and retroviral elements. *BioScience*, **40**, 183–91.

McKenzie, J.A. (1990). Selection at the Dieldrin resistance locus in overwintering populations of *Lucilia cuprina* (Wiedemann). *Australian Journal of Zoology*, **38**, 493–501.

McKenzie, J.A. & Clarke, G.M. (1988). Diazinon resistance, fluctuating asymmetry and fitness in the Australian sheep blowfly, *Lucilia cuprina*. *Genetics*, **120**, 213–20.

McKenzie, J.A. & Yen, J.L. (1995). Genotype, environment and the asymmetry phenotype. Dieldrin-resistance in *Lucilia cuprina* (the Australian sheep blowfly). *Heredity*, **75**, 181–7.

Mearns, L.O., Katz, R.W. & Schneider, S.H. (1984). Extreme high-temperature events: changes in their probabilities with changes in mean temperature. *Journal of Climatology and Applied Meteorology*, **23**, 1601–13.

Merkt, J.R. & Taylor, C.R. (1994). 'Metabolic switch' for desert survival. *Proceedings of the National Academy of Sciences USA*, **91**, 12313–6.

Metz, J.A.J., Nisbet, R.M. & Geritz, S.A.H. (1992). How should we define 'fitness' for general ecological scenarios? *Trends in Ecology and Evolution*, **7**, 198–202.

Milkman, R.D. (1960). The genetic basis of natural variation. II. Analysis of a polygenic system in *Drosophila melanogaster*. *Genetics*, **45**, 377–91.

Miller, P.S. (1994). Is inbreeding depression more severe in a stressful environment? *Zoo Biology*, **13**, 195–208.

Miquel, J., Lundgren, P.R., Bensch, K.G. & Atlan, H. (1976). Effects of temperature on the life span, vitality and fine structure of *Drosophila melanogaster*. *Mechanisms of Ageing and Development*, **5**, 347–70.

Mitchell-Olds, T. & Shaw, R.G. (1987). Regression analysis of natural selection: statistical inference and biological interpretation. *Evolution*, **41**, 1149–61.

Mitton, J.B. (1993). Enzyme heterozygosity, metabolism and developmental stability, *Genetica*, **89**, 47–65.

Monasterio, M. & Sarmiento, L. (1991). Adaptive radiation of *Espeletia* in the cold Andean tropics. *Trends in Ecology and Evolution*, **6**, 387–91.

Moore, L.G. & Regensteiner, J.G. (1983). Adaptation to high altitude. *Annual Review of Anthropology*, **12**, 285–304.

Moritz, C. (1994). Defining 'evolutionary significant units' for conservation. *Trends in Ecology and Evolution*, **9**, 373–5.

Mousseau, T.A. & Dingle, H. (1991). Maternal effects in insect life histories. *Annual Review of Entomology*, **36**, 511–34.

Moya, A., Galiana, A. & Ayala, F.J. (1995). Founder-effect speciation theory: failure of experimental corroboration. *Proceedings of the National Academy of Sciences USA*, **92**, 3983–86.

Mugaas, J.N., Seidensticker, J. & Mahkle-Johnson, K.P. (1993). Metabolic adaptation to climate and distribution of the raccoon *Procyon lotor* and other Procyonidae. *Smithsonian Contributions to Zoology*, **542**, 1–34.

Murphy, G.I. (1968). Pattern in life history and the environment. *American Naturalist*, **102**, 391–403.

Myers, J.H. & Sabath, M.D. (1980). Genetic and phenotypic variability, genetic variance, and the success of establishment of insect introductions for the biological control of weeds. *Proceedings of the V International Symposium on the Biological Control of Weeds*, ed. E.S. Delfosse, pp. 91–102. Melbourne: CSIRO.

Myers, J.P. & Lester, R.T. (1992). Double jeopardy for migrating animals: multiple hits and resource asynchrony. In *Global Warming and Biological Diversity*, ed. R.L. Peters & T.E. Lovejoy, pp. 193–200. New Haven: Yale University Press.

Nevo, E., Filippucci, M.G., Redi,C., Simson, S., Heth, G. & Beiles, A. (1995). Karyotype and genetic evolution in speciation of subterranean mole rats of the genus *Spalax* in Turkey. *Biological Journal of the Linnean Society*, **54**, 203–29.

Neyfakh, A.A. & Hartl, D.L. (1993). Genetic control of the rate of embryonic development: Selection for faster development at elevated temperatures. *Evolution*, **47**, 1625–31.

Nöthel, H. (1987). Adaptation of *Drosophila melanogaster* populations to high mutation pressure: Evolutionary adjustment of mutation rates. *Proceedings of the National Academy of Sciences USA*, **84**, 1045–9.

Orr, H.A. & Coyne, J.A. (1992). The genetics of adaptation: a reassessment. *American Naturalist*, **140**, 725–42.

Osgood, D.W. (1978). Effects of temperature on the development of meristic characters in *Natrix fasciata*. *Copeia*, **1**, 33–47.

Orzack, S.H. & Tuljapurkar, S. (1989). Population dynamics in variable environments. VII. The demography and evolution of iteroparity. *American Naturalist*, **133**, 901–23.

Parsons, P.A. (1973). Genetics of resistance to environmental stresses in *Drosophila* populations. *Annual Review of Genetics*, **7**, 239–65.

Parsons, P.A. (1983a). *The Evolutionary Biology of Colonizing Species*. New York: Cambridge University Press.

Parsons, P.A. (1983b). The genetic basis of quantitative traits: evidence for punctuational evolutionary transitions at the intraspecific level. *Evolutionary Theory*, **6**, 175–84.

Parsons, P.A. (1988). Evolutionary rates: effects of stress upon recombination. *Biological Journal of the Linnean Society*, **35**, 49–68.

Parsons, P.A. (1990). Fluctuating asymmetry: an epigenetic measure of stress. *Biological Reviews*, **63**, 131–145.

Parsons, P.A. (1992a). Biodiversity and climatic change. In *Conservation of Biodiversity for Sustainable Development*, ed. O.T. Sandlund, K. Hindar & A.H.D. Brown, pp. 155–67. Oslo: Scandinavian University Press.

Parsons, P.A. (1992b). Evolutionary adaptation and stress: the fitness gradient. *Evolutionary Biology*, **26**, 191–223.

Parsons, P.A. (1993). Stress, extinctions and evolutionary change: from living organisms to fossils. *Biological Reviews*, **68**, 313–33.

Parsons, P. A. (1994). Habitats, stress, and evolutionary rates. *Journal of evolutionary Biology*, **7**, 387–97.

Parsons, P.A. (1995). Evolutionary response to drought stress: conservation implications. *Biological Conservation*, **74**, 21–7.

Peterson, C.C., Nagy, K.A. & Diamond, J. (1990). Sustained metabolic scope. *Proceedings of the National Academy of Sciences USA*, **87**, 2324–8.

Philippi, T. (1993). Bet-hedging germination of desert annuals: beyond the first year. *American Naturalist*, **142**, 474–87.

Phillippi, T. & Seger, J. (1989). Hedging one's evolutionary bets, revisited. *Trends in Ecology and Evolution*, **4**, 41–4.

Pieri, C., Falasca, M., Recchioni, R., Moroni, F., & Marcheselli, F. (1992). Diet restriction: a tool to prolong the lifespan of experimental animals. Model and currect hypotheses of action. *Comparative Biochemistry and Physiology*, **103A**, 551–4.

Pijanowski, B.C. (1992). A revision of Lack's brood reduction hypothesis. *American Naturalist*, **139**, 1270–92.

Pimm, S.L. (1991). *The Balance of Nature? Ecological Issues in the Conservation of Species and Communities*. Chicago: University of Chicago Press.

Pimm, S.L., Jones, H.L. & Diamond, J. (1988). On the risk of extinction. *American Naturalist*, **132**, 757–85.

Plough, H.H. (1917). The effect of temperature on crossing over in *Drosopila*. *Journal of Experimental Zoology*, **24**, 148–209.

Posthuma, L. & Van Straalen, N.M. (1993). Heavy-metal adaptation in terrestrial invertebrates: a review of occurrence, genetics, physiology and ecological consequences. *Comaprative Biochemistry and Physiology*, **106C**, 11–38.

Price, T.D., Grant, P.R., Gibbs, H.L. & Boag, P.T. (1984). Recurrent patterns of natural selection in a population of Darwin's finches. *Nature*, **309**, 787–9.

Quinn, J.F. & Karr, J.R. (1993). Habitat fragmentation and global change. In *Biotic Interactions and Global Change*, ed. P.M. Kareiva, J.G. Kingsolver & R.B. Huey, pp. 451–63. Sunderland, Massachusetts: Sinauer Associates.

Rankin, M.A. & Burchsted, J.C.A. (1992). The cost of migration in insects. *Annual Review of Entomology*, **37**, 533–59.

Ratner, V.A., Zabanov, S.A., Kolesnikova, O.V. & Vasilyeva, L.A. (1992). Induction of the mobile genetic element *Dm-412* transpositions in the *Drosophila* genome by heat shock treatment. *Proceedings of the National Academy of Sciences USA*, **89**, 5650–4.

Raup, D.M. (1991). *Extinction. Bad Genes or Bad Luck*. New York: W.W. Norton.

Rees, M. (1993). Trade-offs among dispersal strategies in British plants. *Nature*, **366**, 150–2.

Rees, M. (1994). Delayed germination of seeds: A look at the effects of adult longevity, the timing of reproduction, and population age/stage structure. *American Naturalist*, **144**, 43–64.

Rendel, J.M. (1967). *Canalization and Gene Control*. London: Logos Press.

Repasky, P.R. (1991). Temperature and the northern distributions of wintering birds. *Ecology*, **72**, 2274–85.

Rhodes, M.C. & Thayer, C.W. (1991). Mass extinctions: Ecological selectivity and primary production. *Geology*, **19**, 877–80.

Richardson, A.M.M. (1974). Differential climatic selection in natural populations of land snail *Cepaea nemoralis*. *Nature*, **247**, 572–3.

Riska, B., Prout, T. & Turelli, M. (1989). Laboratory estimates of heritability and genetic correlations in nature. *Genetics*, **123**, 865–71.

Robertson, F.W. (1964). The ecological genetics of growth in *Drosophila*. *Genetic Research (Camb.)*, **5**, 107–26.

Rockey, S.J., Hainze, J.H. & Scriber, J.M. (1987). Evidence of a sex-linked diapause response in *Papilio glaucus* subspecies and their hybrids. *Physiological Entomology*, **12**, 181–4.

Roff, D.A. (1991). Life history consequences of bioenergetic and biomechanical constraints on migration. *American Zoologist*, **31**, 205–15.

Roff, D.A. (1992). *The Evolution of Life Histories*. New York: Chapman & Hall.

Rogot, E. & Padgett, S.J. (1976). Associations of coronary and stroke mortality with temperature and snowfall in selected areas of the United States, 1962–1966. *American Journal of Epidemiology*, **103**, 565–75.

Root, T. (1988). Environmental factors associated with avian distributional limits. *Journal of Biogeography*, **15**, 489–505.

Rose, M.R. (1991). *Evolutionary Biology of Aging*. New York: Oxford University Press.

Rose, M.R., Vu, L.N., Park, S.U. & Graves, J.L. (1992). Selection on stress resistance increases longevity in *Drosophila melanogaster*. *Experimental Gerontology*, **27**, 241–50.

Rosenzweig, M.L. & McCord, R.D. (1991). Incumbent replacement: evidence for long-term evolutionary progress. *Paleobiology*, **17**, 202–13.

Roush, R.T. & McKenzie, J.A. (1987). Ecological genetics of insecticide and acaricide resistance. *Annual Review of Entomology*, **32**, 361–80.

Safriel, U.N., Volis, S. & Kark, S. (1994). Core and peripheral populations and global climate change. *Israel Journal of Plant Sciences*, **42**, 331–45.

Schaffer, W.M. (1974). Optimal reproductive effort in fluctuating environments. *American Naturalist*, **108**, 783–90.

Scharloo, W. (1991). Canalization: genetic and developmental aspects. *Annual Review of Ecology and Systematics*, **22**, 65–93.

Schluter, D. & Nychka, D. (1994). Exploring fitness surfaces. *American Naturalist*, **143**, 597–616.

Schmalhausen, I.I. (1949). *Factors of Evolution*, Philadelphia: Blakiston.

Schmidt, K.P. & Levin, D.A. (1985). The comparative demography of reciprocally sown populations of *Phlox drummondii* Hook. I. Survivorships, fecundities, and finite rates of increase. *Evolution*, **39**, 396–404.

Schneider, S.H. (1993). Scenarios of global warming. In *Biotic Interactions and Global Change*, ed. P.M. Kareiva, J.G. Kingsolver & R.B. Huey, pp. 9–23. Sunderland, Massachuetts: Sinauer Associates.

Schoener, T.W. (1983). Field experiments on interspecific competition. *American Naturalist*, **122**, 240–85.

Scott, T.M. & Koehn, R.K. (1990). The effect of environmental stress on the relationship of heterozygosity to growth rate in the coot clam, *Mulinia lateralis* (Say). *Journal of Experimental Marine Biology and Ecology*, **135**, 109–16.

Sepkoski J.J. (1986). Phaneric overview of mass extinction. In *Patterns and Processes in the History of Life*, ed. D.M. Raup & D. Jablonski, pp. 277–95. Berlin: Springer-Verlag.

Service, P.M., Hutchinson, E.W. & Rose, M.R. (1988) Multiple genetic mechanisms for the evolution of senescence in *Drosophila melanogaster*. *Evolution*, **42**, 708–16.

Sheldon, P.R. (1992). Making sense of microevolutionary patterns. In *Evolutionary Patterns and Processes*, ed. D.R. Lees & D. Edwards, pp. 19–31. London: Academic Press.

Shukla, J., Nobre, C. & Sellers, P. (1990). Amazonian deforestation and climate change. *Science*, **247**, 1322–5.

Sibly, R.M. & Calow, P. (1989). A life-cycle theory of response to stress. *Biological Journal of the Linnean Society*, **37**, 101–16.

Smith, P., Townsend, M.G. & Smith, R.H. (1991). A cost of resistance in the brown rat? Reduced growth rate in warfarin-resistant lines. *Functional Ecology*, **5**, 441–7.

Smith, T.B. (1993). Disruptive selection and the genetic basis of bill size polymorphism in the African finch *Pyrenestes*. *Nature*, **363**, 618–20.

Soulé, M.E. & Wilcox, B.A. (1980) *Conservation Biology: An Evolutionary–Ecological Perspective*. Sunderland: Sinauer.

Southwood, T.R.E. (1962). Migration of terrestrial arthropods in relation to habitat. *Biological Reviews*, **37**, 171–214.

Stanley, S.M. (1979). *Macroevolution: Pattern and Process*. San Francisco: W.H. Freeman.

Stanley, S.M., Parsons, P.A., Spence, G.E. & Weber, L. (1980). Resistance of species of the *Drosophila melanogaster* subgroup to environmental extremes. *Australian Journal of Zoology*, **28**, 413–21.

Stearns, S.C. (1992). *The Evolution of Life Histories*. Oxford: Oxford University Press.

Stearns, S.C. & Kawecki, T.J. (1994). Fitness sensitivity and the canalization of life-history traits. *Evolution*, **48**, 1438–50.

Stearns, S.C. & Koella, J.C. (1986). The evolution of phenotypic plasticity in life-history traits: predictions of reaction norms for age and size at maturity. *Evolution*, **40**, 893–913.

Stone, G.N. & Willmer, P.G. (1989). Warm-up rates and body temperatures in bees: the importance of body size, thermal regime and phylogeny. *Journal of Experimental Biology*, **147**, 303–28.

Suchanek, T.H. (1993). Oil impacts on marine invertebrate populations and communities. *American Zoologist*, **33**, 510–23.

Tantawy, A.O. & El-Helw, M.R. (1970). Studies on natural populations of *Drosophila*. II. Heterosis and fitness characters in hybrids between different populations of *Drosophila melanogaster*. *Canadian Journal of Genetics and Cytology*, **12**, 695–710.

Tauber, M.J., Tauber, C.A. & Masaki, S. (1986). *Seasonal Adaptations of Insects*. New York: Oxford University Press.

Templeton, A.R. (1977). Analysis of head shape differences between two interfertile species of Drosophila. *Evolution*, **31**, 630–41.

Templeton, A.R., Hollocher, H. Lawler, S. & Johnston, J.S. (1990). The ecological genetics of abnormal abdomen in *Drosophila mercatorum*. In *Ecological and Evolutionary Genetics of Drosophila*, ed. J.S.F. Barker, W.T. Starmer & R.J. MacIntyre, pp. 17–35. New York: Plenum Press.

Templeton, A.R. & Johnston, J.S. (1988). The measured genotype approach to ecological genetics. In *Population Genetics and Evolution*, ed. G. de Jong, pp. 138–46. Berlin: Springer-Verlag.

Teska, W.R., Smith, M.H. & Novak, J.M. (1990). Food quality, heterozygosity, and fitness correlates in *Peromyscus polionotus*. *Evolution*, **44**, 1318–25.

Thoday, J.M. (1979). Polygene mapping: uses and limitations. In *Quantitative Genetic Variation*, ed. J.N. Thompson, Jr. & J.M. Thoday, pp. 219–233. New York: Academic Press.

Traverse, A. (1988). Plant evolution dances to a different beat: plant and animal evolutionary mechanisms compared. *Historical Biology*, **1**, 277–301.

Tuljapurkar, S.D. (1982). Population dynamics in variable environments. III. Evolutionary dynamics of *r*-selection. *Theoretical Population Biology*, **21**, 141–65.

Ulmasov, K.A., Shammakov, S., Karaev, K. & Evgen'ev, M.B. (1992). Heat shock proteins and thermoresistance in lizards. *Proceedings of the National Academy of Sciences USA*, **89**, 1666–70.

van Noordwijk, A.J., van Balen, J.H. & Scharloo, W. (1988). Heritability of body size in a natural population of the great tit (*Parus major*) and its relation to age and environmental conditions during growth. *Genetical Research*, **51**, 149–62.

Vermeij, G.J. (1987). *Evolution and Escalation: an Ecological History of Life*. Princeton: Princeton University Press.

Vrijenhoek, R.C. & Leberg, P.L. (1991). Let's not throw the baby out with the bathwater: a comment on management for MHC diversity in captive populations. *Conservation Biology*, **5**, 252–4.

Vrijenhoek, R.C, Pfeiler, E. & Wetherington, J.D. (1992). Balancing selection in a desert stream-dwelling fish, *Poeciliopsis monacha*. *Evolution*, **46**, 1642–57.

Waddington, C.H. (1957). *The Strategy of the Genes*. London: George Allen & Unwin.

Ward, P.D. (1992). *On Methuselah's Trail: Living Fossils and Great Extinctions*. New York: W.H. Freeman.

Weiss, S.B., White, R.R., Murphy, D.D. & Ehrlich, P.R. (1987). Growth and dispersal of larvae of the checkerspot butterfly *Euphydryas editha*. *Oikos*, **50**, 161–6.

Westerman, J.M. & Parsons, P.A. (1973). Variation in genetic architecture at different doses of g-radiation as measured by longevity in *Drosophila melanogaster*. *Canadian Journal of Genetics and Cytology*, **15**, 289–98.

White, T.C.R. (1993). *The Inadequate Environment: Nitrogen and the Abundance of Animals*. Berlin: Springer-Verlag.

Wiener, P.A. & Tuljapurkar, S. (1994). Migration in variable environments: Exploring life-history evolution using structured population models. *Journal of Theoretical Biology*, **166**, 75–90.

Williams, T.A. & Nagy, L.M. (1995). Brine shrimp add salt to the stew. *Current Biology*, **5**, 1330–3.

Williamson, P.G. (1981). Palaeontological documentation of speciation in Cenozoic molluscs from Turkana basin. *Nature*, **293**, 437–43.

Wilson, J.B., Ronghua, Y., Mark, A.F. & Agnew, A.D.Q. (1991). A test of the low marginal variance (LMV) theory in *Leptospermum scoparium* (Myrtaceae). *Evolution*, **45**, 780–4.

Windig, J.J. (1994). Genetic correlations and reaction norms in wing pattern of the tropical butterfly *Bicyclus anynana*. *Heredity*, **73**, 459–70.

Wright, S. (1931). Evolution in Mendelian populations. *Genetics*, **16**, 97–159.

Young, T.P. (1990). Evolution of semelparity in Mount Kenya lobelias. *Evolutionary Ecology*, **4**, 157–71.

Zakharov, V.M. (1989). Future prospects for population phenogenetics. *Soviet Science Reviews, Section F, Physiology and General Biology Reviews*, **4**, 1–79.

Zamudio, K.R., Huey, R.B. & Crill, W.D. (1995). Bigger isn't always better: body size, developmental and parental temperature, and territorial success in *Drosophila melanogaster*. *Animal Behaviour*, **49**, 671–7.

Zhuchenko, A.A., Korol, A.B. & Kovtyukh, L.P. (1985). Change of the crossing-over frequency in *Drosophila* during selection for resistance to temperature fluctuations. *Genetica*, **67**, 73–8.

Zotin, A.I. (1990). *Thermodynamic Bases of Biological Processes: Physiological Reactions and Adaptations*. Berlin: Walter de Gruyter.

Index of organisms

Subject index

Printed in the United States
By Bookmasters